AN INTRODUCTION TO INEQUALITIES

NEW MATHEMATICAL LIBRARY

published by
The Mathematical Association of America

The New Mathematical Library (NML) was begun in 1961 by the School Mathematics Study Group to make available to high school students short expository books on various topics not usually covered in the high school syllabus. In a decade the NML matured into a steadily growing series of some twenty titles of interest not only to the originally intended audience, but to college students and teachers at all levels. Previously published by Random House and L. W. Singer, the NML became a publication series of the Mathematical Association of America (MAA) in 1975. Under the auspices of the MAA the NML will continue to grow and will remain dedicated to its original and expanded purposes.

AN INTRODUCTION
TO INEQUALITIES

by

Edwin Beckenbach

University of California, Los Angeles

Richard Bellman

The RAND Corporation

3

THE MATHEMATICAL ASSOCIATION
OF AMERICA

Illustrations by Carl Bass

Thirteenth Printing

Library of Congress Catalog Card Number: 61-6228

Complete Set ISBN-0-88385-600-X
Vol. 3   0-88385-603-4

Manufactured in the United States of America

To
Alan, Barbara, Edwin,
Eric, Kirstie, Lenann,
Suzann, and Tommy

NEW MATHEMATICAL LIBRARY

1 Numbers: Rational and Irrational *by Ivan Niven*
2 What is Calculus About? *by W. W. Sawyer*
3 An Introduction to Inequalities *by E. F. Beckenbach and R. Bellman*
4 Geometric Inequalities *by N. D. Kazarinoff*
5 The Contest Problem Book I Annual High School Examinations 1950–1960. Complied and with solutions *by Charles T. Salkind*
6 The Lore of Large Numbers *by P. J. Davis*
7 Uses of Infinity *by Leo Zippin*
8 Geometric Transformations I *by I. M. Yaglom, translated by A. Shields*
9 Continued Fractions *by Carl D. Olds*
10 Graphs and Their Uses *by Oystein Ore*
11 Hungarian Problem Books I and II, Based on the Eötvös
12 Competitions 1894–1905 and 1906-1928 *translated by E. Rapaport*
13 Episodes from the Early History of Mathematics *by A. Aaboe*
14 Groups and Their Graphs *by I. Grossman and W. Magnus*
15 The Mathematics of Choice *by Ivan Niven*
16 From Pythagoras to Einstein *by K. O. Friedrichs*
17 The Contest Problem Book II Annual High School Examinations 1961–1965. Complied and with solutions *by Charles T. Salkind*
18 First Concepts of Topology *by W. G. Chinn and N. E. Steenrod*
19 Geometry Revisited *by H. S. M. Coxeter and S. L. Greitzer*
20 Invitation to Number Theory *by Oystein Ore*
21 Geometric Transformations II *by I. M. Yaglom, translated by A. Shields*
22 Elementary Cryptanalysis—A Mathematical Approach *by A. Sinkov*
23 Ingenuity in Mathematics *by Ross Honsberger*
24 Geometric Transformations III *by I. M. Yaglom, translated by A. Shenitzer*
25 The Contest Problem Book III Annual High School Examination 1966–1972. Complied and with solutions *by C. T. Salkind and J. M. Earl*
26 Mathematical Methods in Science *by George Pólya*
27 International Mathematical Olympiads 1959–1977. Compiled and with solutions *by Samuel L. Greitzer*
28 The Mathematics of Games and Gambling *by Edward W. Packel*
Others titles in preparation

Note to the Reader

This book is one of a series written by professional mathematicians in order to make some important mathematical ideas interesting and understandable to a large audience of high school students and laymen. Most of the volumes in the *New Mathematical Library* cover topics not usually included in the high school curriculum; they vary in difficulty, and, even within a single book, some parts require a greater degree of concentration than others. Thus, while the reader needs little technical knowledge to understand most of these books, he will have to make an intellectual effort.

If the reader has so far encountered mathematics only in classroom work, he should keep in mind that a book on mathematics cannot be read quickly. Nor must he expect to understand all parts of the book on first reading. He should feel free to skip complicated parts and return to them later; often an argument will be clarified by a subsequent remark. On the other hand, sections containing thoroughly familiar material may be read very quickly.

The best way to learn mathematics is to *do* mathematics, and each book includes problems, some of which may require considerable thought. The reader is urged to acquire the habit of reading with paper and pencil in hand; in this way mathematics will become increasingly meaningful to him.

The authors and editorial committee are interested in reactions to the books in this series and hope that readers will write to: Anneli Lax, Editor, New Mathematical Library, NEW YORK UNIVERSITY, THE COURANT INSTITUTE OF MATHEMATICAL SCIENCES, 251 Mercer Street, New York, N. Y. 10012.

The Editors

CONTENTS

Note to the Reader		vii
Preface		3
Chapter 1	Fundamentals	5
Chapter 2	Tools	15
Chapter 3	Absolute Value	25
Chapter 4	The Classical Inequalities	47
Chapter 5	Maximization and Minimization Problems	79
Chapter 6	Properties of Distance	99
Symbols		113
Answers to Exercises		115
Index		131

AN INTRODUCTION TO INEQUALITIES

Preface

Mathematics has been called the science of tautology; that is to say, mathematicians have been accused of spending their time proving that things are equal to themselves. This statement (appropriately by a philosopher) is rather inaccurate on two counts. In the first place, mathematics, although the language of science, is not a science. Rather, it is a creative art. Secondly, the fundamental results of mathematics are often *inequalities* rather than *equalities*.

In the pages that follow, we have presented three aspects of the theory of inequalities. First, in Chapters 1, 2, and 3, we have the axiomatic aspect. Secondly, in Chapter 4, we use the products of the preceding chapters to derive the basic inequalities of analysis, results that are used over and over by the practicing mathematician. In Chapter 5, we show how to use these results to derive a number of interesting and important maximum and minimum properties of the elementary symmetric figures of geometry: the square, cube, equilateral triangle, and so on. Finally, in Chapter 6, some properties of distance are studied and some unusual distance functions are exhibited.

There is thus something for many tastes, material that may be read consecutively or separately. Some readers will want to understand the axiomatic approach that is basic to higher mathematics.

They will enjoy the first three chapters. In addition, in Chapter 3 there are many illuminating graphs associated with inequalities. Other readers will prefer for the moment to take these results for granted and turn immediately to the more analytic results. They will find Chapter 4 to their taste. There will be some who are interested in the many ways in which the elementary inequalities can be used to solve problems that ordinarily are treated by means of calculus. Chapter 5 is intended for these. Readers interested in generalizing notions and results will enjoy the analysis of some strange non-Euclidean distances described in Chapter 6.

Those whose appetites have been whetted by the material presented here will want to read the classic work on the subject, *Inequalities,* by G. H. Hardy, J. E. Littlewood, and G. Pólya, Cambridge University Press, London, 1934. A more recent work containing different types of results is *Inequalities,* by E. F. Beckenbach and R. Bellman, *Ergebnisse der Mathematik,* Julius Springer Verlag, Berlin, 1961.

E. F. B.
R. B.

Santa Monica, California, 1960

CHAPTER ONE

Fundamentals

1.1 The "Greater-than" Relationship

You will recall that the symbol ">" means "greater than" or "is greater than." Then you can readily answer the question: Is $3 > 2$? Of course it is.

But is $-3 > -2$? Admittedly, -3 is a "greater negative number" than -2, but this statement does not answer what is meant by the question. If the real numbers (zero and the positive and negative rational and irrational numbers) are represented geometrically in the usual way by points on a horizontal number scale directed to the right, as indicated in Fig. 1.1, then the numbers appear in order of

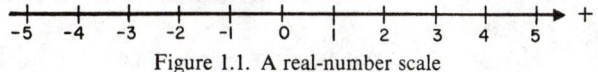

Figure 1.1. A real-number scale

increasing value from left to right. The point representing -2 appears *to the right* of the point representing -3, and accordingly $-2 > -3$. Similarly,

$$(1.1) \quad 4 > -4, \quad 3 > 2, \quad 0 > -2, \quad -1 > -2, \quad 1 > 0.$$

Hence we have the following geometric rule for determining inequality: *Let a and b be any two real numbers represented by points on a horizontal number scale directed to the right. Then $a > b$ if and*

5

only if the point representing a lies to the right of the point representing b.

You can *say* that $-3 > -2$, or that $-300 > -2$, but it is not true according to the foregoing geometric rule.

In dealing with inequalities, it is often more fruitful and even necessary to work algebraically instead of graphically. The geometric rule given above has, in terms of the *basic notion of positive number,* the following simple algebraic equivalent:

DEFINITION. Let a and b be any two real numbers. Then $a > b$ if and only if $a - b$ is a positive number.

Thus, if $a = -2$ and $b = -3$, then $a - b = -2 - (-3) = 1$ is positive. Hence $-2 > -3$, as noted above in the geometric discussion. You might check the inequalities in (1.1) by the present algebraic method of subtraction and verify each of the following inequalities both by the geometric method and by the algebraic method:

$$\pi > 3, \quad 2 > 0, \quad 1 > -9, \quad \sqrt{2} > 1, \quad -\frac{1}{2} > -40.$$

1.2 The Sets of Positive Numbers, Negative Numbers, and Zero

You will note that in the preceding section we defined the inequality $a > b$ in terms of *positive numbers.* The set P of positive numbers, and similarly the set N of negative numbers, as well as the special set O having as its only member the number 0, play essential roles in the study of inequalities. In fact, while of course we shall freely use the familiar algebraic (field) properties of the real number system, such as the commutative, associative, and distributive laws, a basic thesis of this entire tract is that *all order relationships in the real number system —all algebraic inequalities—can be made to rest on two simple axioms regarding the set P of positive numbers.* These axioms will be presented in the following section.

Symbolically, for "*a* is positive" we write "*a* ε *P,*" read in full as "*a* is a *member* (or *element*) of the set *P.*" Thus we have 5 ε *P*, 0 ε *O*, -3 ε *N*.

Let us look briefly at the foregoing sets, *P, N,* and *O,* and their members.

The number *zero,* of course, is the unique member 0 of the set *O;* it satisfies the equation

$$a + 0 = a$$

for any real number *a.*

Regarding the set N of negative numbers, it is important to distinguish the idea of *the negative of a number* from the idea of *a negative number:*

The negative of a number a is defined to be the number $-a$ such that

$$(a) + (-a) = 0.$$

Thus, if $a = -3$ then the negative of a is $-(-3) = 3$, since $(-3) + (3) = 0$. Similarly, if $a = 0$ then $-a = 0$ since $0 + 0 = 0$.

A negative number is defined to be the negative of a positive number. Thus, you recognize 3, 1/2, 9/5, π, $\sqrt{2}$ as being members of the set P of positive numbers. Then -3, $-1/2$, $-9/5$, $-\pi$, $-\sqrt{2}$ are members of the set N of negative numbers.

We shall not attempt to define the basic notion of a positive number, but shall now proceed to characterize these numbers by means of two basic axioms.

1.3 The Basic Inequality Axioms

The following simple propositions involving the set P of positive numbers are stated without proof; accordingly, they are called *axioms*. It is interesting to note that they are the *only* propositions needed, along with the familiar algebraic structure of the real number system,† for the development of the entire theory of inequalities.

AXIOM I. *If a is a real number, then one and only one of the following statements is true: a is the unique member* 0 *of the set O; a is a member of the set P of positive numbers;* $-a$ *is a member of the set P.*

AXIOM II. *If a and b are members of the set P of positive numbers, then the sum* $a + b$ *and the product ab are members of the set P.*

The three alternatives listed in Axiom I relate an arbitrary real number a and its negative $-a$ as follows: If a is *zero,* then $-a$ is *zero,* as already noted; if a is *positive,* then $-a$ is *negative* by the foregoing definition of negative number; and if $-a$ is *positive,* then $a = -(-a)$ must be *negative,* again by the definition of negative number. Thus a and $-a$ are paired in the sets P, N, and O as indicated in Table 1.

———

† But see the footnote on page 12.

TABLE 1. Pairings of Numbers and Their Negatives

Number	Set		
a	P	N	O
$-a$	N	P	O

In the geometric representation (Fig. 1.1), the points representing a and $-a$ either coincide at the point representing 0 or lie on opposite sides of that point.

1.4 Reformulation of Axiom I

Axiom I is concerned with the set P of positive numbers, and the inequality $a > b$ was defined in terms of the set P. Let us reformulate this axiom in terms of the inequality relationship.

If a and b are arbitrary real numbers, then their difference, $a - b$, is a real number; accordingly, Axiom I can be applied to $a - b$. Thus either $(a - b)\,\varepsilon\,O$ (that is, $a = b$), or $(a - b)\,\varepsilon\,P$ (that is, $a > b$), or $-(a - b) = (b - a)\,\varepsilon\,P$ (that is, $b > a$), and these three possibilities are mutually exclusive. Hence, the following statement is a consequence of Axiom I:

AXIOM I′. *If a and b are real numbers, then one and only one of the following relationships holds:*

$$a = b, \qquad a > b, \qquad b > a.$$

In particular, Axiom I′ asserts, in the special case $b = 0$, that if a is a real number then exactly one of the following alternatives holds: $a = 0$ (that is, $a\,\varepsilon\,O$), or $a > 0$ (that is, $a\,\varepsilon\,P$), or $0 > a$ (that is, $-a\,\varepsilon\,P$). Accordingly, Axiom I can be deduced from Axiom I′.

If a statement S can be deduced from—i.e., is a consequence of— a statement T, we say "T implies S." We have just seen that Axiom I implies Axiom I′ and also that Axiom I′ implies Axiom I. If each of two statements implies the other, we say that they are equivalent. Thus, Axioms I and I′ are equivalent.

To illustrate Axioms I and I′, consider the numbers $a_1 = 3$, $a_2 = -4$, $b_1 = 0$, $b_2 = 3$.

Illustrating Axiom I, you note that $a_1\,\varepsilon\,P$, $-a_2\,\varepsilon\,P$, $b_1\,\varepsilon\,O$, and $b_2\,\varepsilon\,P$; you note also that $a_1\,\not\varepsilon\,O$ (read "a_1 is not a member of the set O"), and $-a_1\,\not\varepsilon\,P$, etc.

Illustrating Axiom I', you have

$$a_1 - b_1 = 3 - 0 = 3, \qquad a_1 - b_1 > 0, \qquad a_1 > b_1;$$
$$a_1 - b_2 = 3 - 3 = 0, \qquad a_1 - b_2 = 0, \qquad a_1 = b_2;$$
$$a_2 - b_1 = -4 - 0 = -4, \qquad b_1 - a_2 > 0, \qquad b_1 > a_2;$$
$$a_2 - b_2 = -4 - 3 = -7, \qquad b_2 - a_2 > 0, \qquad b_2 > a_2.$$

You note, then, that in each of the four instances one and only one of the three relationships given in Axiom I' holds. This illustration of Axiom I' will be continued in the following section, as additional inequality relationships are introduced.

1.5 Additional Inequality Relationships

In place of an inequality such as $b > a$, you might equally well write $a < b$, read "a is less than b." The two inequalities are entirely equivalent and neither is generally preferable to the other. In the foregoing illustration of Axiom I', the sign ">" was used throughout for the sake of consistency. But you might just as well have considered it preferable to be consistent in writing the a's before the b's in all the relationships. Then you would have had

$$(1.2) \qquad a_1 > b_1, \qquad a_1 = b_2, \qquad a_2 < b_1, \qquad a_2 < b_2.$$

Likewise,

$$-4 < 4, \qquad 2 < 3, \qquad -2 < 0, \qquad -2 < -1, \qquad 0 < 1,$$
$$3 < \pi, \qquad 0 < 2, \qquad -9 < 1, \qquad 1 < \sqrt{2}, \quad -40 < -\frac{1}{2}.$$

The symbols ">" and "<" represent *strict* inequalities.

Two other relationships considered in the study of inequalities are the *mixed* inequalities $a \geq b$ and $a \leq b$, read "a is greater than or equal to b" and "a is less than or equal to b," respectively. The first of these, $a \geq b$, means that either $a > b$ or $a = b$; for example, $3 \geq 2$ and also $2 \geq 2$. The second, $a \leq b$, means that either $a < b$ or $a = b$; thus $1 \leq 2$ and also $2 \leq 2$.

In (1.2) it is stated that *one* of the three relationships listed in Axiom I' holds in each instance. But the axiom itself asserts further that *only one* of the relationships holds. Therefore, to make the illustration of Axiom I' complete, you really should add the statements

$$(1.3) \qquad a_1 \nleq b_1, \qquad a_1 \nleqq b_2, \qquad a_2 \ngeqq b_1, \qquad a_2 \ngeqq b_2,$$

read "a_1 is neither less than nor equal to b_1," etc.

You naturally feel that these negative statements in (1.3) are superfluous, and indeed no one would claim that you should ordinarily write them down to complete the information contained in (1.2). This is due to the fact that the exclusiveness principle—the "one and only one" aspect—of Axiom I or I′ is taken for granted.

Because of the exclusiveness principle, the respective relationships in (1.2) and (1.3) clearly are equivalent; that is, each implies the other. Nevertheless, the negation of an inequality is often a very useful concept.

If you tend to confuse the two symbols ">" and "<", you might note that in a valid inequality, such as $3 > 2$ or $2 < 3$, the *larger* (open) end of the symbol is toward the *greater* number while the *smaller* (pointed) end is toward the *lesser* number.

1.6 Products Involving Negative Numbers

What sort of number is the product of a positive number and a negative number? Or the product of two negative numbers? We can use Axioms I and II and some of their consequences to determine the answers to these questions.

If $a \, \varepsilon \, P$ and $b \, \varepsilon \, N$, then $-b \, \varepsilon \, P$ according to Table 1, so that the product $a(-b) \, \varepsilon \, P$ by Axiom II. Hence $-[a(-b)] \, \varepsilon \, N$ by the definition of a negative number; but $-[a(-b)] = ab$ by the usual algebraic rules for interchanging parentheses and minus signs:

$$-[a(-b)] = -[-(ab)] = ab.$$

Therefore $ab \, \varepsilon \, N$, and thus we have the following result:

THEOREM 1.1. *The product ab of a positive number a and a negative number b is a negative number.*

Similarly, if $a \, \varepsilon \, N$ and $b \, \varepsilon \, N$, then $-a \, \varepsilon \, P$ and $-b \, \varepsilon \, P$ according to Table 1. Hence, by Axiom II, their product $(-a)(-b) \, \varepsilon \, P$. But by the rules of algebra, $(-a)(-b) = ab$, and therefore $ab \, \varepsilon \, P$. Hence we get this result:

THEOREM 1.2. *The product ab of two negative numbers a and b is a positive number.*

In particular, by this last result and Axiom II, the square of any real number other than zero is a positive number. Of course, $0^2 = 0$.

Thus we obtain one of the simplest and most useful results in the entire theory of inequalities:

THEOREM 1.3. *Any real number a satisfies the inequality $a^2 \geq 0$. The sign of equality holds if and only if $a = 0$.*

1.7 "Positive" and "Negative" Numbers

By this time, you may realize the strength of Axioms I and II. You might be amused to learn that from them you could even determine which of the nonzero real numbers belong to the set P of positive numbers and which belong to the set N of negative numbers—as if you didn't already know!

To see this, let us for the moment put "positive" and "negative" in quotation marks to indicate information obtained from the axioms.

Let us start with the number $a = 1$. Since $a \neq 0$, it follows from Theorem 1.3 that $a^2 > 0$. Thus a^2 is "positive." But

$$a^2 = 1^2 = 1,$$

so that 1 is "positive."

Next, let us try $a = 2$. Since we have now determined that 1 is "positive," since $1 + 1 = 2$, and since by Axiom II the sum of two "positive" numbers is "positive," it follows that 2 is "positive."

Now let $a = \frac{1}{2}$; then $2a = 1$. Thus the product of the "positive" number 2 and the number a is the "positive" number 1. But if a were "negative," then the product of 2 and a would be "negative," by Theorem 1.1. Therefore $a = \frac{1}{2}$ must be "positive."

Thus the numbers 1, 2, $\frac{1}{2}$ are "positive" and hence, by Table 1, the numbers -1, -2, $-\frac{1}{2}$ are "negative."

Continuing, we can show that the integers 3, 4, etc.; the fractions $\frac{1}{3}$, $\frac{1}{4}$, etc.; and the fractions $\frac{2}{3}$, $\frac{4}{3}$, $\frac{3}{4}$, $\frac{5}{4}$, etc., are "positive," and accordingly that -3, -4, $-\frac{1}{3}$, etc., are "negative." Thus, for any nonzero rational number we can determine whether it is "positive" or "negative."

Finally, the limiting processes that are used in defining irrational numbers can be applied to determine, from our knowledge as to which rational numbers are "positive" and which are "negative," whether a given irrational number is "positive" or "negative" in the completely ordered field of real numbers.† We shall not discuss irrational numbers in any detail in this book; for an interesting treatment of them, you should read *Numbers: Rational and Irrational,* by Ivan Niven, also in this *New Mathematical Library* series.

Exercises

1. Make a sketch showing points representing the following numbers on a horizontal number scale directed to the right:

$$3, \quad -1, \quad 0, \quad -1.5, \quad \pi - 3, \quad 3 - \pi, \quad \sqrt{2}, \quad 2, \quad -2, \quad -3.$$

   Rewrite the numbers in increasing order, giving the result as a continued inequality of the form $a < b < c$, etc.

2. Put a stroke through ε, thus $\not\varepsilon$, if the statement is false:

   (a) $-3 \, \varepsilon \, N$, (f) $a^2 \, \varepsilon \, N$,

   (b) $0 \, \varepsilon \, P$, (g) $(a^2 + 1) \, \varepsilon \, P$,

   (c) $5 \, \varepsilon \, O$, (h) $-2^2 \, \varepsilon \, P$,

   (d) $\sqrt{2} \, \varepsilon \, N$, (i) $(a^2 + 1) \, \varepsilon \, O$,

   (e) $(\pi - 3) \, \varepsilon \, P$, (j) $-3 \, \varepsilon \, P$.

3. Fill in each blank with P, N, or O so that a true statement results:

   (a) $\dfrac{48}{273} - \dfrac{49}{273} \, \varepsilon$ _____, (f) $7^2 - 4(2)(6) \, \varepsilon$ _____,

   (b) $\dfrac{721}{837} - \dfrac{721}{838} \, \varepsilon$ _____, (g) $93(72 + \frac{1}{2}) - 93(72) \, \varepsilon$ _____,

   (c) $\dfrac{-23}{32} - \dfrac{-25}{32} \, \varepsilon$ _____, (h) $93(72 - \frac{1}{2}) - 93(72) \, \varepsilon$ _____,

   (d) $\dfrac{-23}{32} - \dfrac{-23}{33} \, \varepsilon$ _____, (i) $\dfrac{2+3}{4+5} - \frac{1}{2}\left(\frac{2}{4} + \frac{3}{5}\right) \, \varepsilon$ _____,

   (e) $\dfrac{-1}{-2} - \dfrac{1}{-2} \, \varepsilon$ _____, (j) $(-3)^2 - 3^2 \, \varepsilon$ _____.

† Here *ordered* means that Axioms I and II are satisfied, and *complete* refers to the basic property that, if a non-null set of real numbers has an upper bound, then it has a least upper bound. For example, the set $\{1, 1.4, 1.41, \ldots\}$ of rational approximations to $\sqrt{2}$ is bounded from above by 2, or by 1.5; hence it has a least upper bound (which we denote by $\sqrt{2}$). The corresponding point on the number scale (see page 5) divides that scale into a left-hand and a right-hand portion. Since there is at least one "positive" rational number—e.g. 1 or 1.4—in the left-hand portion, we say that $\sqrt{2}$ is "positive." In Sec. 1.7 we have shown that the rational numbers can be *ordered* in *only one* way, and we have stated that, similarly, the real numbers can be ordered in only one way. The property of complete ordering, or its equivalent, is used in defining the real numbers in terms of the rational numbers, and accordingly it is taken as a postulate for the real numbers rather than as an Axiom III for inequalities.

4. Fill in each blank with $>$, $<$, or $=$ so that a true statement results:

(a) $\dfrac{48}{273}$ ____ $\dfrac{49}{273}$,

(f) 7^2 ____ $4(2)(6)$,

(b) $\dfrac{721}{837}$ ____ $\dfrac{721}{838}$,

(g) $93(72 + \frac{1}{2})$ ____ $93(72)$,

(c) $\dfrac{-23}{32}$ ____ $\dfrac{-25}{32}$,

(h) $93(72 - \frac{1}{2})$ ____ $93(72)$,

(d) $\dfrac{-23}{32}$ ____ $\dfrac{-23}{33}$,

(i) $\dfrac{2+3}{4+5}$ ____ $\dfrac{1}{2}\left(\dfrac{2}{4} + \dfrac{3}{5}\right)$,

(e) $\dfrac{-1}{-2}$ ____ $\dfrac{1}{-2}$,

(j) $(-3)^2$ ____ 3^2.

5. Label T for true or F for false:

(a) $-2 \geq -3$ ____,

(f) $-1 \leq 2$ ____,

(b) $0 \leq 0$ ____,

(g) $\frac{3}{2} < \frac{3}{4}$ ____,

(c) $0 > -1$ ____,

(h) $-\frac{2}{3} \geq -\frac{3}{3}$ ____,

(d) $\frac{1}{2} < \frac{1}{3}$ ____,

(i) $1 - 2^2 < -2^2$ ____,

(e) $-\frac{1}{2} < -\frac{1}{3}$ ____,

(j) $1 < 0$ ____.

6. Write the negative of each of the following numbers:

$$-2, \quad 3 - \pi, \quad (3 - \pi)^2, \quad \frac{a}{b - c}, \quad 0, \quad \sqrt{b^2 - 4ac}.$$

7. Fill in the blanks to give affirmative relationships equivalent to the stated negative relationships:

(a) $a \not< b$, a ____ b,

(d) $a \not\leq b$, a ____ b,

(b) $a \neq b$, a ____ b,

(e) $a \not> b$, a ____ b,

(c) $a \not> b$, a ____ b,

(f) $a \not\geq b$, a ____ b.

8. Show that each positive number p is greater than each negative number n.

9. For two real numbers a and b, what single conclusion can you draw if you can show that $a \geq b$ and that $a \leq b$?

10. Prove by mathematical induction from Axiom II that if $a_1, a_2, \ldots, a_n$ are positive, then the sum $a_1 + a_2 + \cdots + a_n$ and the product $a_1 a_2 \cdots a_n$ are positive. (See the comments in Sec. 2.6.)

11. Use Axioms I and II to show that $\frac{2}{3}$ is a "positive" number.

CHAPTER TWO

Tools

2.1 Introduction

In dealing with inequalities, the only basic assumptions that are ultimately used are the two axioms discussed in Chapter 1, along with the real number system and its laws, such as the distributive law, mathematical induction, etc. Still, there are several simple theorems that derive from these axioms and that occur so frequently in the development and application of the theory that they might almost be called "tools of the trade."

These theorems, or operational rules, and their proofs are attractive and interesting on their own merits. Moreover, they furnish an excellent illustration of the way mathematicians build up an elaborate system of results from a few significant basic notions and assumptions. The proofs are usually short, but nevertheless complete; and in just a few places they call for the touch of ingenuity that makes mathematics the fascinating subject that it is.

In the present chapter, some of these theorems are listed, illustrated, and proved. The letters a, b, c, etc., that occur in the statement of the theorems will be understood to represent real numbers, arbitrary except for explicitly given constraints.

For convenience, the theorems, or rules as we shall sometimes call them, will be stated here only for the "$>$" case. In each instance, there is an equivalent "$<$" rule. Thus, to the "$>$" rule for transitivity, "If $a > b$ and $b > c$, then $a > c$," there corresponds the equivalent "$<$" rule, "If $a < b$ and $b < c$, then $a < c$."

You can similarly construct the "$<$" rule that is the companion to each of the other "$>$" rules given in this chapter, but watch out for mathematical pitfalls! If a rule involves a positive multiplier, say $c > 0$, and requires that the difference of two arbitrary quantities be positive, then the companion "$<$" rule still has $c > 0$ (or $0 < c$ if you prefer), not $c < 0$. Thus the companion rule to "If $a > b$ and $c > 0$, then $ac > bc$" is "If $a < b$ and $c > 0$, then $ac < bc$."

The statement of the theorem at the beginning of each section is divided into two paragraphs. The simpler first paragraph, dealing with the strict "$>$" inequality sign, contains the heart of the result. The second paragraph involves the mixed "$\geq$" inequality sign, and in addition it sometimes includes the case of an arbitrary number n of real values. Thus it deals with a more general and inclusive situation. The proof is usually given for the more inclusive general case only, but it can readily be specialized to apply to the first paragraph.

The illustrations of the theorems given in this chapter will sometimes involve the stated "$>$" rule, and sometimes the implied "$<$" rule.

2.2 Transitivity

THEOREM 2.1. *If $a > b$ and $b > c$, then $a > c$.*
More generally, if $a_1 \geq a_2, a_2 \geq a_3, \ldots, a_{n-1} \geq a_n$, then $a_1 \geq a_n$, with $a_1 = a_n$ if and only if all the a's are equal.

Thus, if a consideration of your personal expenditures has led you to observe that you spend more money on Saturday than on any weekday, and that you spend at least as much on Sunday as on Saturday, then you can conclude that you spend more money on Sunday than you do on any weekday.

Again, the solution of Exercise 1 in Chapter 1 is

$$-3 < -2 < -1.5 < -1 < 3 - \pi < 0 < \pi - 3 < \sqrt{2} < 2 < 3;$$

this might be interpreted narrowly as meaning only that each of the first nine members of the set is less than the *immediately* succeeding number, thus: $-3 < -2$, $-2 < -1.5$, etc. If so, it still *implies*, by the foregoing transitivity rule, that each of these numbers is less than any succeeding number; for example, $-3 < -1.5$, $-1 < 2$, $3 - \pi < \sqrt{2}$.

PROOF. The transitivity rule could be proved by simple mathematical induction (for which see the comments in Sec. 2.6). But for this first rule, we shall give a single direct proof involving four real numbers.

Suppose, then, that $a_1 \geq a_2$, $a_2 \geq a_3$, $a_3 \geq a_4$. By the algebraic definition of inequality, each of the quantities $a_1 - a_2$, $a_2 - a_3$, $a_3 - a_4$ is either in the set P or in the set O. Therefore the sum

$$(a_1 - a_2) + (a_2 - a_3) + (a_3 - a_4) = a_1 - a_4$$

is either in P, by Axiom II, or in O; and it is in O if and only if $a_1 - a_2 = 0$, $a_2 - a_3 = 0$, $a_3 - a_4 = 0$. Thus $a_1 \geq a_4$, and the sign of equality holds if and only if $a_1 = a_2$, $a_2 = a_3$, $a_3 = a_4$.

The proof for the general case is left as an exercise.

2.3 Addition

THEOREM 2.2. *If $a > b$ and $c > d$, then $a + c > b + d$. If $a > b$, and c is any real number, then $a + c > b + c$.*

More generally, if $a_1 \geq b_1, a_2 \geq b_2, \ldots, a_n \geq b_n$, then

$$(2.1) \qquad a_1 + a_2 + \cdots + a_n \geq b_1 + b_2 + \cdots + b_n.$$

The sign of equality holds in (2.1) if and only if $a_1 = b_1$, $a_2 = b_2, \ldots, a_n = b_n$.

Thus, if the inequalities $1 < \sqrt{2}$ and $3 < \pi$ are added, they yield $1 + 3 < \sqrt{2} + \pi$. This last inequality combined with $-1 = -1$ gives $3 < \sqrt{2} + \pi - 1$. And all five of these relationships added together produce $10 < 3(\sqrt{2} + \pi) - 2$.

PROOF. As in the case of transitivity, an inductive proof is again available. This time, however, a direct proof of the general case will be given. Since by hypothesis each of the quantities $a_1 - b_1$, $a_2 - b_2, \ldots, a_n - b_n \, \varepsilon \, P$ or $\varepsilon \, O$, then by the generalization of Axiom II given in Exercise 10 of Chapter 1, the sum $(a_1 - b_1) + (a_2 - b_2) + \cdots + (a_n - b_n) = (a_1 + a_2 + \cdots + a_n) - (b_1 + b_2 + \cdots + b_n) \, \varepsilon \, P$, unless it $\varepsilon \, O$ with $a_1 - b_1 = 0$, $a_2 - b_2 = 0, \ldots, a_n - b_n = 0$. Thus

$$a_1 + a_2 + \cdots + a_n \geq b_1 + b_2 + \cdots + b_n,$$

and equality holds if and only if $a_1 = b_1$, $a_2 = b_2, \ldots, a_n = b_n$.

2.4 Multiplication by a Number

THEOREM 2.3. *If $a > b$ and $c > 0$, then $ac > bc$. If $a > b$ and $c < 0$, then $ac < bc$.*

More generally, if $a \geq b$ and $c > 0$, then $ac \geq bc$, with $ac = bc$ if and only if $a = b$. If $a \geq b$ and $c < 0$, then $ac \leq bc$, with $ac = bc$ if and only if $a = b$.

Thus, multiplying the terms of an inequality by a positive number leaves the sign of inequality unchanged, but multiplying by a *negative* number *reverses* it. In particular, for $c = -1$, if $a \geq b$ then $-a \leq -b$.

For example, if you multiply $3 > 2$ by 1 and -1, you obtain $3 > 2$ and $-3 < -2$, respectively.

PROOF. It is given that $a \geq b$, so that $a - b \varepsilon P$ or εO. If $c \varepsilon P$, then it follows from Axiom II that $c(a - b) = ca - cb \varepsilon P$ or εO; that is, $ca \geq cb$.

But if $c \varepsilon N$, then it follows from Theorem 1.1 that $c(a - b) \varepsilon N$ or εO; hence, $-[c(a - b)] = cb - ca \varepsilon P$ or εO, so that $cb \geq ca$.

In either case, the sign of equality holds if and only if $a = b$.

2.5 Subtraction

THEOREM 2.4. *If $a > b$ and $c > d$, then $a - d > b - c$. If $a > b$, and c is any real number, then $a - c > b - c$.*

More generally, if $a \geq b$ and $c \geq d$, then $a - d \geq b - c$, with $a - d = b - c$ if and only if $a = b$ and $c = d$.

Note that d is subtracted from a, and c from b—*not* c from a, or d from b.

Thus, by subtraction, the inequalities $7 > 6$ and $5 > 3$ yield $7 - 3 > 6 - 5$, that is, $4 > 1$; but the inequality $7 - 5 > 6 - 3$ is false. Or, to illustrate the rule in terms of "$<$", the inequalities $-5 < 10$ and $-4 < -3$ yield $-5 - (-3) < 10 - (-4)$; that is, $-2 < 14$.

PROOF. Applying the rule for multiplying an inequality by a negative number (Theorem 2.3) to $c \geq d$, we get $-c \leq -d$, that is, $-d \geq -c$, where the sign of equality holds if and only if $c = d$. Now applying the rule for adding inequalities to $a \geq b$ and $-d \geq -c$, we obtain $a + (-d) \geq b + (-c)$, or $a - d \geq b - c$, with $a - d = b - c$ if and only if $a = b$ and $c = d$.

Exercises

1. Show that if $a < b$ then $a < \frac{1}{2}(a + b) < b$.
2. Show that

$$(a^2 - b^2)(c^2 - d^2) \leq (ac - bd)^2$$

and

$$(a^2 + b^2)(c^2 + d^2) \geq (ac + bd)^2$$

   for all $a, b, c, d,$ and that the signs of equality hold if and only if $ad = bc$.
3. Show that

$$(a^2 - b^2)^2 \geq 4ab(a - b)^2$$

   for all $a, b,$ and that the sign of equality holds if and only if $a = b$.
4. Prove the general ">" transitivity rule by mathematical induction.

2.6 Multiplication

THEOREM 2.5. *If $a > b > 0$ and $c > d > 0$, then $ac > bd$.*
More generally, if $a_1 \geq b_1 > 0, a_2 \geq b_2 > 0, \ldots, a_n \geq b_n > 0$, then

(2.2) $$a_1 a_2 \cdots a_n \geq b_1 b_2 \cdots b_n.$$

The sign of equality holds in (2.2) if and only if $a_1 = b_1, a_2 = b_2, \ldots, a_n = b_n$.

Thus, from $2 > 1$ and $4 > 3$ you obtain $(2)(4) > (1)(3)$, or $8 > 3$. But note that $-1 > -2$ and $-3 > -4$, yet $(-1)(-3) < (-2)(-4)$; accordingly, the requirement that the numbers be positive cannot be dropped.

PROOF. We shall give a proof by mathematical induction. The standard procedure in constructing such a proof consists of the following steps. First, the statement to be proved for all positive integers n is tested for the first one or two; then, under the assumption that the statement is true for all integers up to and including a certain one, say $k - 1$, it is proved that the statement is true also for the next integer, k. Since k can be any integer > 1 (in particular, let $k - 1 = 1$ or 2, for which the statement was verified), we may conclude that the statement is indeed true for all positive integers.

For $n = 1$, the conclusion $a_1 \geq b_1$ of Theorem 2.5 is simply a repetition of the hypothesis. This observation is sufficient for the first step in the proof by mathematical induction, but we shall give the proof also for $n = 2$; that is, we shall show that if $a_1 \geq b_1 > 0$ and $a_2 \geq b_2 > 0$, then $a_1 a_2 \geq b_1 b_2$.

The inequality

$$(2.3) \qquad\qquad a_1 a_2 \geq b_1 a_2$$

follows from the rule for the multiplication of an inequality by a positive number; the sign of equality holds if and only if $a_1 = b_1$. The inequality

$$(2.4) \qquad\qquad b_1 a_2 \geq b_1 b_2$$

follows from the same rule; the sign of equality holds if and only if $a_2 = b_2$. Now, the desired inequality

$$(2.5) \qquad\qquad a_1 a_2 \geq b_1 b_2$$

follows from (2.3) and (2.4) by the transitivity rule (Theorem 2.1); the sign of equality holds in (2.5) if and only if it holds in (2.3) and (2.4), that is, if and only if $a_1 = b_1$ and $a_2 = b_2$.

We have demonstrated that the inequality (2.2) holds for $n = 1$ and $n = 2$.

Assume that the inequality (2.2) is true for $n = 1, 2, \ldots, k - 1$, that is, for products of $k - 1$ numbers:

$$(2.6) \qquad\qquad a_1 a_2 \cdots a_{k-1} \geq b_1 b_2 \cdots b_{k-1};$$

equality holds if and only if $a_1 = b_1$, $a_2 = b_2$, $\ldots$, $a_{k-1} = b_{k-1}$. Then from the rule for multiplication by a positive number (Theorem 2.3), when we multiply the inequality (2.6) by a_k we obtain

$$(2.7) \qquad\qquad (a_1 a_2 \cdots a_{k-1}) a_k \geq (b_1 b_2 \cdots b_{k-1}) a_k;$$

the sign of equality holds if and only if

$$a_1 a_2 \cdots a_{k-1} = b_1 b_2 \cdots b_{k-1}.$$

Also, from the same rule, when we multiply the inequality

$$a_k \geq b_k$$

by $b_1 b_2 \cdots b_{k-1}$, we get

$$(2.8) \qquad\qquad (b_1 b_2 \cdots b_{k-1}) a_k \geq (b_1 b_2 \cdots b_{k-1}) b_k;$$

the sign of equality holds if and only if $a_k = b_k$. It now follows from (2.7) and (2.8), by the transitivity rule, that

$$a_1 a_2 \cdots a_{k-1} a_k \geq b_1 b_2 \cdots b_{k-1} b_k;$$

equality holds if and only if $a_1 = b_1$, $a_2 = b_2$, $\ldots$, $a_k = b_k$.

2.7 Division

THEOREM 2.6. *If* $a > b > 0$ *and* $c > d > 0$, *then* $a/d > b/c$. *In particular, for* $a = b = 1$, *if* $c > d > 0$, *then* $1/d > 1/c$.

More generally, if $a \geq b > 0$ *and* $c \geq d > 0$, *then* $a/d \geq b/c$, *with* $a/d = b/c$ *if and only if* $a = b$ *and* $c = d$.

Note that a is divided by d, and b by c—*not* a by c, or b by d.

Thus, by division, the inequalities $7 > 6$ and $5 > 3$ yield $7/3 > 6/5$; but the inequality $7/5 > 6/3$ is false.

The two given inequalities also yield $1/6 > 1/7$ and $1/3 > 1/5$.

PROOF. We have

$$\frac{a}{d} - \frac{b}{c} = \frac{ac - bd}{cd}.$$

The denominator $cd \, \varepsilon \, P$ by Axiom II, since $c \, \varepsilon \, P$ and $d \, \varepsilon \, P$. Also, since $a \geq b$ and $c \geq d$, it follows from Theorem 2.5 that $ac \geq bd$; hence the numerator $ac - bd \, \varepsilon \, P$ or $\varepsilon \, O$, and actually $ac - bd \, \varepsilon \, P$ unless $a = b$ and $c = d$. By Theorem 1.1, the product of a negative number and a positive number is negative; but the product

$$cd \left(\frac{ac - bd}{cd} \right)$$

is equal to the nonnegative number $ac - bd$, and cd is positive. Hence

$$\frac{ac - bd}{cd} = \frac{a}{d} - \frac{b}{c}$$

is nonnegative. Thus $a/d \geq b/c$, with $a/d = b/c$ if and only if $a = b$ and $c = d$.

Exercises

1. From the inequality

$$\left(\frac{1}{\sqrt{a}} - \frac{1}{\sqrt{b}} \right)^2 \geq 0,$$

show that

$$\frac{2}{(1/a) + (1/b)} \leq \sqrt{ab}$$

for all positive a, b. Under what circumstances does the sign of equality hold?

2. Show that the sum of a positive number and its reciprocal is at least 2; that is, show that

$$a + \frac{1}{a} \geq 2$$

for all positive values a. For what value of a does the sign of equality hold?

3. Show that

$$a^2 + b^2 + c^2 \geq ab + bc + ca$$

for all a, b, c.

4. Show that

$$(a^2 - b^2)(a^4 - b^4) \leq (a^3 - b^3)^2$$

and

$$(a^2 + b^2)(a^4 + b^4) \geq (a^3 + b^3)^2$$

for all a, b.

5. Show that

$$a^2b + b^2c + c^2a + ab^2 + bc^2 + ca^2 \geq 6abc$$

for all nonnegative a, b, c.

6. Show that

$$(a^2 - b^2)^2 \geq (a - b)^4$$

for all a, b satisfying $ab \geq 0$ and that

$$(a^2 - b^2)^2 \leq (a - b)^4$$

for all a, b satisfying $ab \leq 0$.

7. Show that

$$a^3 + b^3 \geq a^2b + ab^2$$

for all a, b satisfying $a + b \geq 0$.

8. For any of Exercises 3 through 7 that you have worked, determine under what circumstances the signs of equality hold.

2.8 Powers and Roots

THEOREM 2.7. *If* $a > b > 0$, *if* m *and* n *are positive integers, and if* $a^{1/n}$ *and* $b^{1/n}$ *denote positive nth roots, then*

$$a^{m/n} > b^{m/n} \qquad and \qquad b^{-m/n} > a^{-m/n}.$$

More generally, if $a \geq b > 0$, *if* m *is a nonnegative integer and* n *a positive integer, and if* $a^{1/n}$ *and* $b^{1/n}$ *denote positive nth roots, then*

$$(2.9) \qquad a^{m/n} \geq b^{m/n} \qquad and \qquad b^{-m/n} \geq a^{-m/n},$$

with $a^{m/n} = b^{m/n}$ *and* $b^{-m/n} = a^{-m/n}$ *if and only if either (i)* $a = b$ *or (ii)* $m = 0$.

For $a = 9$ and $b = 4$, certain values of $a^{m/n}$, $b^{m/n}$, $b^{-m/n}$, $a^{-m/n}$ appear in Table 2. For each positive value of m/n you will note that $9^{m/n} > 4^{m/n}$, while $4^{-m/n} > 9^{-m/n}$.

TABLE 2. Samples of Powers of Numbers

$\dfrac{m}{n}$	$9^{m/n}$	$4^{m/n}$	$4^{-m/n}$	$9^{-m/n}$
0	1	1	1	1
$\frac{1}{2}$	3	2	$\frac{1}{2}$	$\frac{1}{3}$
1	9	4	$\frac{1}{4}$	$\frac{1}{9}$
$\frac{3}{2}$	27	8	$\frac{1}{8}$	$\frac{1}{27}$
2	81	16	$\frac{1}{16}$	$\frac{1}{81}$

PROOF. If $m = 0$, then $a^{m/n} = b^{m/n} = b^{-m/n} = a^{-m/n} = 1$, so that the sign of equality holds in (2.9) in this case.

If $m \neq 0$, then $a^m \geq b^m$ by the rule for multiplication of inequalities (Theorem 2.5); the sign of equality holds if and only if $a = b$. If it were true that $a^{1/n} < b^{1/n}$, then it would also be true that $(a^{1/n})^n < (b^{1/n})^n$, or $a < b$; but, by hypothesis, $a \geq b$. Accordingly, $a^{1/n} \geq b^{1/n}$. Therefore $a^{m/n} \geq b^{m/n}$, with $a^{m/n} = b^{m/n}$ if and only if $a = b$.

For negative exponents, let

$$a^{m/n} = c, \qquad b^{m/n} = d.$$

Then

$$a^{-m/n} = \frac{1}{c}, \qquad b^{-m/n} = \frac{1}{d}.$$

Since we have just shown that

$$c \geq d,$$

it follows from Theorem 2.6 that

$$\frac{1}{d} \geq \frac{1}{c};$$

that is,

$$b^{-m/n} \geq a^{-m/n},$$

the sign of equality holding if and only if $c = d$, that is, $a = b$.

The rule can be extended to positive and negative irrational powers.

Exercises

1. Show that
$$\frac{a+b}{2} \le \left(\frac{a^2+b^2}{2}\right)^{1/2}$$
for all a, b. Under what circumstance does the sign of equality hold?

2. Show that if a, b, and c, d are positive (and c and d are rational), then
$$(a^c - b^c)(a^d - b^d) \ge 0$$
and
$$a^{c+d} + b^{c+d} \ge a^c b^d + a^d b^c.$$
Under what circumstance do the signs of equality hold?

3. To what does the second inequality in Exercise 2 reduce in the case $c = d = 1$? in the case $c = d = \frac{1}{2}$?

4. For $bd > 0$, show that $a/b \le c/d$ if and only if $ad \le bc$, and that the sign of equality holds in each place if and only if it does in the other.

5. Show that if $a/b \le c/d$, then
$$\frac{a+b}{b} \le \frac{c+d}{d},$$
and that the sign of equality holds if and only if $ad = bc$.

6. Show that if $a/b \le c/d$, with a, b, c, d positive, then
$$\frac{a}{a+b} \le \frac{c}{c+d},$$
and that the sign of equality holds if and only if $ad = bc$.

7. Show that if $a/b \le c/d$, with b and d positive, then
$$\frac{a}{b} \le \frac{a+c}{b+d} \le \frac{c}{d},$$
and that the sign of equality holds if and only if $ad = bc$.

8. Verify, by the test given in Exercise 4, that the four inequalities in the conclusion of Exercises 5, 6, and 7 are valid for the values $a = 2, b = 3$, $c = 5, d = 6$.

9. Write the "$<$" inequality rule equivalent to the first paragraph of the "$>$" inequality rule for transitivity, addition, multiplication by a number, subtraction, multiplication, division, and powers and roots.

CHAPTER THREE

Absolute Value

3.1 Introduction

In Chapter 1, as you will recall, the inequality $a > b$ was defined in terms of the set P of positive numbers. You may also recall that for the validity of several of the results of Chapter 2, such as Theorem 2.5 concerning the multiplication of inequalities, it was necessary to specify that certain of the numbers involved should be positive. Again, in many instances the fractional powers of numbers that appear in Theorem 2.7 would not even be real if the numbers themselves were negative; consider, for instance, $a^{1/2}$ with $a = -9$. Many of the fundamental inequalities, which will be derived in Chapter 4, involve just such fractional powers of numbers. It is natural, then, that we should often restrict our attention to positive numbers or to nonnegative numbers (positive numbers and zero) in this study.

In applied problems involving inequalities, we often deal with weights, volumes, etc., and with the *magnitudes,* or *absolute values,* of certain mathematical objects such as real numbers, complex numbers, vectors. The magnitudes of all these are measured by nonnegative numbers. Thus, even though you may choose to denote gains by positive numbers and losses by negative numbers, a loss of \$3 is still a loss of greater magnitude than a loss of \$2; the absolute value of -3 is greater than the absolute value of -2.

In this chapter we shall define and study some properties of the absolute value of real numbers, for application to inequalities in subsequent chapters. We shall also exhibit the graphs of some interesting but rather "off-beat" functions involving absolute values and shall present some new ideas regarding them.

3.2 Definition

The *absolute value* of the real number a is denoted by $|a|$; it can be defined in a variety of equivalent ways. We shall consider several of the definitions here.

DEFINITION. The *absolute value* $|a|$ of the real number a is defined to be a if a is positive or zero, and to be $-a$ if a is negative.

Thus, $|2| = 2$, $|0| = 0$, and $|-2| = -(-2) = 2$.

The principal disadvantage of the foregoing definition is that it is unsuitable for algebraic manipulation. Thus (see Theorem 3.2 later in this chapter) for all a, b it is true that

$$|a + b| \leq |a| + |b|,$$

as you can verify by considering separately the cases in which a and b are both positive, one positive and the other negative, both negative, one zero and one positive, one zero and one negative, and both zero. But it would be preferable to give a unified proof of such a result by standard algebraic procedures; this will be done in Sec. 3.8, after a different but equivalent definition of absolute value in terms of squares and square roots will have been given.

We could rephrase the above definition somewhat differently:

The *absolute value* $|a|$ of the real number a is 0 if $a \, \varepsilon \, O$, and otherwise $|a|$ is the positive member of the set $\{a, -a\}$.

Thus, if $a = 2$, then $|a|$ is the positive member of $\{2, -2\}$, i.e., 2; if $a = -2$, then $|a|$ is the positive member of $\{-2, -(-2)\}$, i.e., 2. But this characterization of $|a|$ has the same algebraic disadvantages as the preceding one.

3.3 Special Symbols

The next two characterizations of $|a|$ depend on two useful special symbols, max $\{\ \}$ and $\{\ \}^+$, which we shall now define.

For any set $\{a_1, a_2, \ldots, a_n\}$ of real numbers, the symbol max $\{a_1, a_2, \ldots, a_n\}$ *denotes the greatest member of the set.*

If there are only one or two members in the set, we still say "greatest" in this connection; and if the greatest value is taken on by more than one of the members, then any one of these is understood to be the greatest. Thus,

$$\max \{3, 7, 0, -2, 5\} = 7, \quad \max \{4, 4\} = 4, \quad \max \{-3, -1\} = -1.$$

With some difficulty, arithmetic operations can be performed on expressions involving the symbol max { }; for example,

$$\frac{(\max \{4, -3\})(\max \{0, 5\}) + \max \{-4, 4\} - \max \{-9, -8\}}{2 \max \{1, 4\}} = 4.$$

In particular, consider max $\{a, -a\}$; if $a = 2$, then

$$\max \{a, -a\} = \max \{2, -2\} = 2 = |a|;$$

if $a = -3$, then

$$\max \{a, -a\} = \max \{-3, -(-3)\} = 3 = |a|;$$

if $a = 0$, then

$$\max \{a, -a\} = \max \{0, 0\} = 0 = |a|;$$

and so on. Thus, for all a,

(3.1) $$\max \{a, -a\} = |a|,$$

so that (3.1) *gives us another characterization of* $|a|$.

Let us turn now to the consideration of a second special symbol.

The symbol $\{a_1, a_2, \ldots, a_n\}^+$ *denotes the greatest member of the set* $\{a_1, a_2, \ldots, a_n\}$ *if there is at least one nonnegative member of the set; but if all the members of the set are negative then the symbol denotes* 0.

Thus

$$\{3, 7, 0, -2, 5\}^+ = 7, \qquad \{4, 4\}^+ = 4, \qquad \{-3, -1\}^+ = 0.$$

As in the case of max { }, it is awkward but possible to deal arithmetically with expressions involving the symbol { }$^+$; for example,

$$\frac{(\{4, -3\}^+)(\{0, 5\}^+) + \{-4, 4\}^+ - \{-9, -8\}^+}{2\{1, 4\}^+} = 3.$$

The symbols max { } and { }$^+$ are not equivalent, as the above examples show; in fact, you can easily see from the definitions that

$$\{a_1, a_2, \ldots, a_n\}^+ = \max \{0, a_1, a_2, \ldots, a_n\}$$
$$= \max\left\{0, \max \{a_1, a_2, \ldots, a_n\}\right\}.$$

Thus it follows that

$$\{a_1, a_2, \ldots, a_n\}^+ \geq \max \{a_1, a_2, \ldots, a_n\},$$

where the sign of equality holds if and only if there is at least one nonnegative member of the set $\{a_1, a_2, \ldots, a_n\}$.

Accordingly, since the special set $\{a, -a\}$ does have a nonnegative member for all a,

$$\{a, -a\}^+ = \max \{a, -a\} = |a|.$$

Thus, the equation

$$\{a, -a\}^+ = |a|$$

might also be considered as a definition of $|a|$.

Exercises

1. Determine the values of
   (a) max $\{-7, -4, -1\}$,
   (b) max $\{3, \pi, \sqrt{2}\}$,
   (c) max $\{-7, 0, -1\}$,
   (d) max $\{0, 4, 1\}$,
   (e) max $\{3, -3, 3\}$,

   (f) $\{-7, -4, -1\}^+$,
   (g) $\{3, \pi, \sqrt{2}\}^+$,
   (h) $\{-7, 0, -1\}^+$,
   (i) $\{0, 4, 1\}^+$,
   (j) $\{3, -3, 3\}^+$.

2. For any set $\{a_1, a_2, \ldots, a_n\}$ of real values, the symbol min $\{a_1, a_2, \ldots, a_n\}$ denotes the least member of the set $\{a_1, a_2, \ldots, a_n\}$, and the symbol $\{a_1, a_2, \ldots, a_n\}^-$ denotes the least member of the set $\{0, a_1, a_2, \ldots, a_n\}$. Determine the values of
   (a) min $\{-7, -4, -1\}$,
   (b) min $\{3, \pi, \sqrt{2}\}$,
   (c) min $\{-7, 0, -1\}$,
   (d) min $\{0, 4, 1\}$,
   (e) min $\{3, -3, 3\}$,

   (f) $\{-7, -4, -1\}^-$,
   (g) $\{3, \pi, \sqrt{2}\}^-$,
   (h) $\{-7, 0, -1\}^-$,
   (i) $\{0, 4, 1\}^-$,
   (j) $\{3, -3, 3\}^-$.

3. Determine the value of

$$(\max \{-1, -2\})(\{-1, -2\}^+) - (\min \{1, 2\})(\{1, 2\}^-).$$

4. Show that

$$\max \left\{ \max \{a, b, c\}, \max \{d, e\} \right\} = \max \{a, b, c, d, e\}.$$

5. Give an example showing that the inequality

$$\max \{a, b\} + \max \{c, d\} \geq \max \{a, b, c, d\}$$

is not always valid.

6. Show that

$$\{a, b\}^+ + \{c, d\}^+ \geq \{a, b, c, d\}^+.$$

7. Show that

$$\{a_1, a_2, \ldots, a_n\}^+ \geq \max \{a_1, a_2, \ldots, a_n\}$$
$$\geq \min \{a_1, a_2, \ldots, a_n\}$$
$$\geq \{a_1, a_2, \ldots, a_n\}^-.$$

Is there any set for which the strict inequality sign holds in all three places?

8. Show that if $a = \max \{a, b, c\}$, then $-a = \min \{-a, -b, -c\}$.

9. Show that $\{-a, -b\}^- = -\{a, b\}^+$.

10. Show that $\max \{a_1, a_2, \ldots, a_n\} = \max \left\{a_1, \max \{a_2, a_3, \ldots, a_n\}\right\}$.

3.4 Graphical Considerations

A graphical representation can furnish a vivid and striking picture of the behavior of a function, whether we are dealing with mean daily temperatures, the fluctuations of the stock market, $|x|$, or what-not; for one thing, a graph lets us see at a glance some of the over-all properties of the function that otherwise might have been obscure.

For example, the symbols max { } and { }$^+$ are made more meaningful by a consideration of the graphs of

$$y = \max \left\{\frac{1}{2}x - \frac{1}{2}, \ -\frac{3}{4}x - \frac{7}{4}\right\}$$

and

$$y = \left\{\frac{1}{2}x - \frac{1}{2}, \ -\frac{3}{4}x - \frac{7}{4}\right\}^+,$$

which are shown in Figs. 3.1 and 3.2, respectively. In these figures, the graphs of

$$y = \frac{1}{2}x - \frac{1}{2} \quad \text{and} \quad y = -\frac{3}{4}x - \frac{7}{4}$$

are extended as dashed lines.

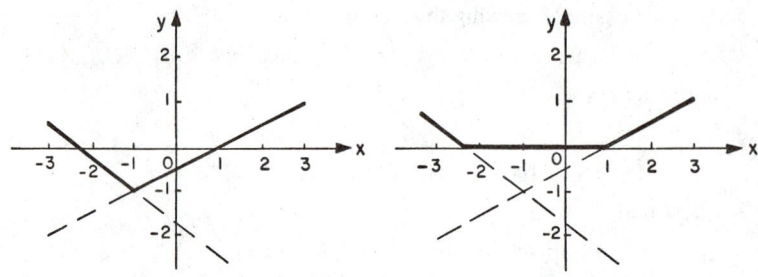

Figure 3.1.

Graph of $y = \max\left\{\frac{1}{2}x - \frac{1}{2}, \ -\frac{3}{4}x - \frac{7}{4}\right\}$, $-3 \leq x \leq 3$

Figure 3.2.

Graph of $y = \left\{\frac{1}{2}x - \frac{1}{2}, \ -\frac{3}{4}x - \frac{7}{4}\right\}^+$, $-3 \leq x \leq 3$

Let us now construct the graph of the function defined by the equation $y = |x|$. *This graph provides us with a visual characterization of absolute value.* It will be sufficient for our purposes to restrict our attention to the incomplete graph corresponding to the interval $-3 \leq x \leq 3$.

In the construction of this graph, it is helpful and interesting to consider first the graph of $y' = x$, that is, the graph of the set of ordered pairs of real numbers (x, y') such that $y' = x$, and also the graph of $y'' = -x$, as shown in Figs. 3.3 and 3.4, respectively.

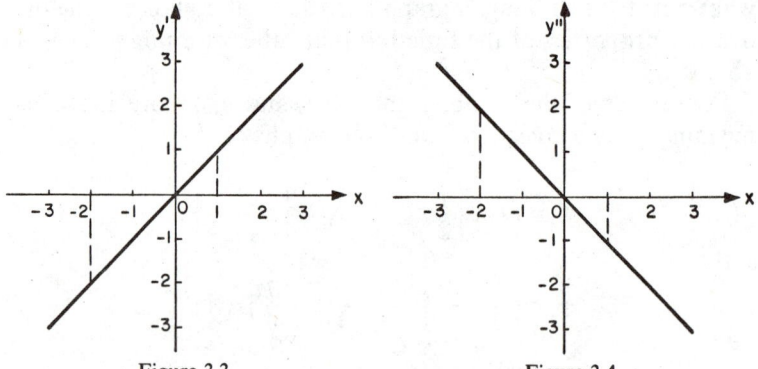

Figure 3.3.

Graph of $y' = x$, $-3 \leq x \leq 3$

Figure 3.4.

Graph of $y'' = -x$, $-3 \leq x \leq 3$

From these figures and the definition

$$|x| = \max \{x, -x\} = \max \{y', y''\},$$

you can see at once that the graph of $y = |x|$ is simply the graph of $y = \max \{y', y''\}$, as illustrated in Fig. 3.5. Thus for each abscissa x, the greater of the ordinates y' and y'' was chosen from Figs. 3.3 and 3.4 as the ordinate y in Fig. 3.5. For example, when $x = -2$, the greater ordinate is $y'' = 2$; when $x = 1$, the greater ordinate is $y' = 1$, etc.

Figure 3.6 shows the graph of $y = -|x|$.

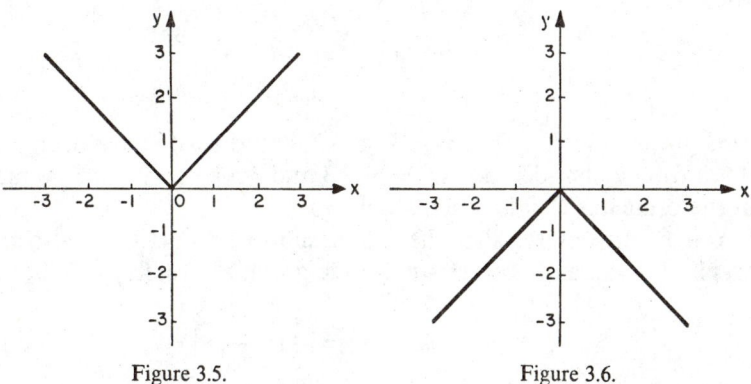

Figure 3.5.
Graph of $y = |x|$, $-3 \leq x \leq 3$

Figure 3.6.
Graph of $y = -|x|$, $-3 \leq x \leq 3$

Looking at the four graphs in Figs. 3.3 through 3.6, you will note that for no value of the abscissa x is any one of the corresponding ordinates less than $-|x|$ or more than $|x|$. Explicitly, from Figs. 3.3, 3.5, and 3.6, you can read off the following result, which of course you *might* have discovered and proved without considering the graphs at all:

THEOREM 3.1. *For each real number a,*

$$-|a| \leq a \leq |a|.$$

The first sign of equality holds if and only if $a \leq 0$, and the second if and only if $a \geq 0$.

Theorem 3.1 follows, for instance, from the fact that $a = -|a|$ if $a \, \varepsilon \, N$ or $a \, \varepsilon \, O$, and $a = |a|$ if $a \, \varepsilon \, P$ or $a \, \varepsilon \, O$, and from the fact (see Exercise 8 in Chapter 1) that any positive number is greater than any negative number.

Now, for exercises, let us consider the graphs of some more complicated functions involving absolute values.

Consider first the graph of

$$y = \frac{1}{2}(x + |x|).$$

For $x \geq 0$, $|x| = x$, and hence

$$y = \frac{1}{2}(x + x) = x;$$

but for $x < 0$, $|x| = -x$, so that

$$y = \frac{1}{2}(x - x) = 0.$$

This graph, which is shown in Fig. 3.7, could easily have been pictured from a consideration of Figs. 3.3 and 3.5 by taking the average of the ordinates y', y for each abscissa x.

You might observe that the graph shown in Fig. 3.7 is also the graph of $y = \max \{0, x\}$, as well as the graph of $y = \{x\}^+$. Thus,

$$\{x\}^+ = \max \{0, x\} = \frac{1}{2}(x + |x|)$$

for all x.

Now let us look at the graph of

(3.2) $$y = 2|x + 1| + |x| + |x - 1| - 3$$

in the interval $-2 \leq x \leq 2$. For $1 \leq x$, the terms in the right-hand member of (3.2) can be written thus:

$$2|x + 1| = 2x + 2, \quad |x| = x, \quad |x - 1| = x - 1, \quad -3 = -3,$$

so that for $1 \leq x$ we have

$$y = 2x + 2 + x + x - 1 - 3 = 4x - 2.$$

For $0 < x < 1$, the first two members in the right-hand member of (3.2) can be written as before, but

$$|x - 1| = 1 - x, \quad \text{not} \quad |x - 1| = x - 1.$$

(Can you explain why?) Accordingly, for $0 < x < 1$,

$$y = 2x + 2 + x + 1 - x - 3 = 2x.$$

Similarly, for $-1 \le x \le 0$,

$$y = 2x + 2 - x + 1 - x - 3 = 0;$$

and for $x < -1$,

$$y = -2x - 2 - x + 1 - x - 3 = -4x - 4.$$

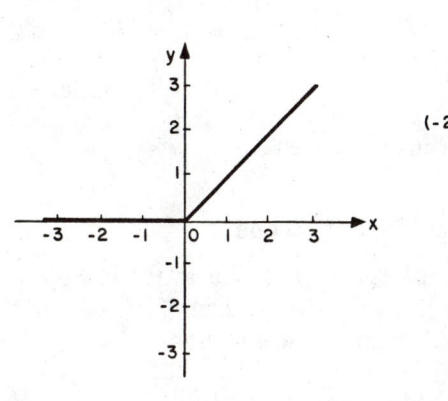

Figure 3.7.

Graph of $y = \frac{1}{2}(x + |x|)$, $-3 \le x \le 3$

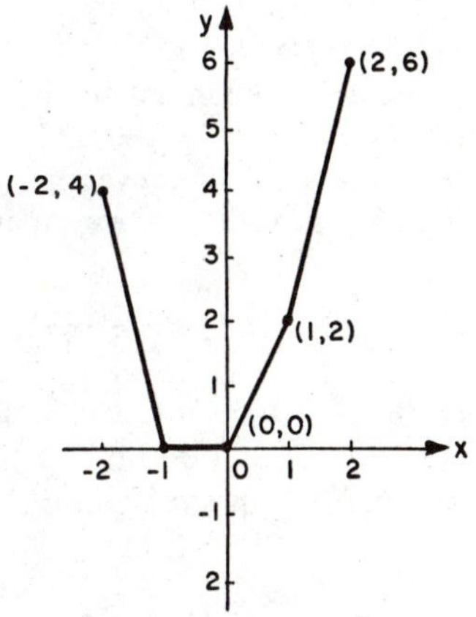

Figure 3.8. Graph of

$$y = 2|x + 1| + |x| + |x - 1| - 3,$$
$$-2 \le x \le 2$$

Thus eq. (3.2) is equivalent to a different linear equation in each of the foregoing intervals. Plotting the corresponding linear segments in the appropriate intervals, we obtain the continuous graph shown in Fig. 3.8.

Exercises

1. From a consideration of Figs. 3.4, 3.5, 3.6, obtain a result for $-a$ analogous to Theorem 3.1 for a.

2. For $-3 \le x \le 3$, sketch the graphs of

   (a) $y = \frac{1}{2}(x - |x|)$,   (b) $y = \frac{1}{2}(|x| - x)$,   (c) $y = -\frac{1}{2}(x + |x|)$.

3. Determine which of the graphs in Exercise 2 is also the graph of

   (d) $y = \min\{0, x\}$,   (g) $y = \{x\}^-$,
   (e) $y = \max\{0, -x\}$,   (h) $y = \{-x\}^+$,
   (f) $y = \min\{0, -x\}$,   (i) $y = \{-x\}^-$.

4. For $-3 \leq x \leq 3$, sketch the graph of

    (a) $y = \max \{x, -x - 2\}$,   (c) $y = \{x, -x - 2\}^+$,

    (b) $y = \min \{x, -x - 2\}$,   (d) $y = \{x, -x - 2\}^-$.

5. For $-3 \leq x \leq 3$, sketch the graph of

$$y = 2|x - 1| - |x| + 2|x + 1| - 5.$$

6. A function f defined by an equation $y = f(x)$ is said to be *even* if $f(-x) = f(x)$ for all x, and is said to be *odd* if $f(-x) = -f(x)$ for all x. Thus, the function defined by $y = x^2$ is even since $(-x)^2 = x^2$, and the function defined by $y = x^3$ is odd since $(-x)^3 = -x^3$; of course, some functions are neither even nor odd. Which of the functions illustrated in Figs. 3.3 through 3.6 are even and which are odd?

3.5 The "Sign" Function

Another function closely related to $|x|$ is illustrated in Fig. 3.9. It is the function denoted by $y = \text{sgn } x$ (read "sign of x," but not to be confused with sin x), and defined by the equations

(3.3)
$$
\begin{aligned}
\text{sgn } x &= +1 &\text{for} &&\quad x > 0, \\
\text{sgn } x &= 0 &\text{for} &&\quad x = 0, \\
\text{sgn } x &= -1 &\text{for} &&\quad x < 0.
\end{aligned}
$$

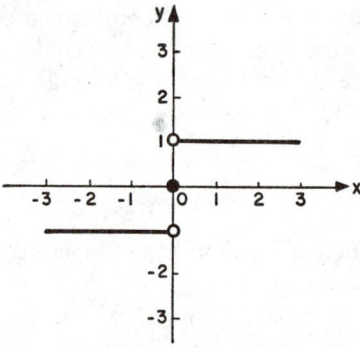

Figure 3.9. Graph of $y = \text{sgn } x$, $-3 \leq x \leq 3$

In the graph, the points corresponding to $(0, 1)$ and $(0, -1)$ are enclosed in unshaded, or open, circles to emphasize the fact that they are *not* included in the graph, and the point corresponding to $(0, 0)$ is covered by a shaded circle to emphasize that it *is* included.

The function $y = \operatorname{sgn} x$ is related to $y = |x|$ through the notion of *slope*, which is defined as follows:

Let L be a nonvertical line in the coordinate plane, and let $P_1:(x_1, y_1)$ and $P_2:(x_2, y_2)$ be distinct points on L; see Figs. 3.10 and 3.11. In going from P_1 to P_2, the vertical *rise* is the *directed* distance $y_2 - y_1$, and the horizontal *run* is the *directed* distance $x_2 - x_1 \neq 0$. Of course, either the rise or the run or both might be negative; thus in Fig. 3.11, the rise $y_2 - y_1$ is negative (actually, then, a fall). The ratio of the rise to the run, which has the same value for all pairs of distinct points P_1, P_2 on L, is defined to be the *slope m of L*:

$$m = \frac{y_2 - y_1}{x_2 - x_1}.$$

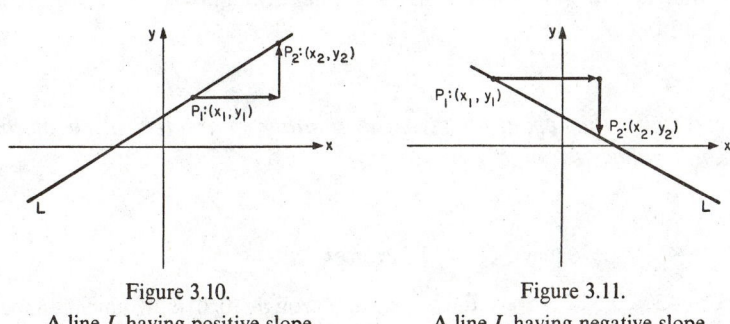

Figure 3.10.	Figure 3.11.
A line L having positive slope	A line L having negative slope

It should be easy for you to verify that the slopes of the linear graphs of $y = x$ and $y = -x$ are 1 and -1, respectively. See Figs. 3.3 and 3.4.

Now consider the *slope* of the graph of $y = |x|$, shown in Fig. 3.5; at the same time, keep in mind the *ordinate* of the graph of $y = \operatorname{sgn} x$ shown in Fig. 3.9.

For $x > 0$, the graph of $y = |x|$ coincides with the linear graph of $y = x$, and it has *slope $m = 1$*. Also, for $x > 0$, the graph of $y = \operatorname{sgn} x$ has *ordinate $y = 1$*.

For $x < 0$, the graph of $y = |x|$ coincides with the linear graph of $y = -x$, and it has *slope $m = -1$*. Also, for $x < 0$, the graph of $y = \operatorname{sgn} x$ has *ordinate $y = -1$*.

For $x = 0$, the slope of the graph of $y = |x|$ is undefined, but you can say that the *right-hand slope* at $(0, 0)$ is 1 and the *left-hand slope* at that point is -1. The *average* of these slopes is $\frac{1}{2}[1 + (-1)] = 0$. Also, for $x = 0$, the graph of $y = \operatorname{sgn} x$ has *ordinate $y = 0$*.

Thus, $y = |x|$ and $y = \text{sgn}\, x$ are related geometrically as follows:

For $x \neq 0$, the value of $y = \text{sgn}\, x$ is equal to the value of the slope of the graph $y = |x|$; and for $x = 0$, it is equal to the average of the values of the right-hand and left-hand slopes.

It is interesting to note that although we have been analyzing two relatively simple functions, $y = |x|$ and $y = \text{sgn}\, x$, the graph of $y = |x|$ is somewhat peculiar in that it does not have a continuous slope; and the graph of $y = \text{sgn}\, x$ is even more strange in that it is, itself, discontinuous. We shall not here attempt to define continuity and discontinuity, but the meaning should be intuitively clear in the present instances.

The function defined by $y = \text{sgn}\, x$ is related to $y = |x|$ in a second instructive way. A moment's reflection shows that, for each real number a,

$$a\,\text{sgn}\, a = |a|;$$

accordingly, *this equation gives us another characterization of the absolute value $|a|$ of a number a.*

Exercises

1. For $-3 \leq x \leq 3$, draw lines passing through $(0,0)$ and having slopes (a) $m = 0$, (b) $m = \frac{2}{3}$, (c) $m = -1$.

2. Sketch the portions of the graphs of

   (a) $y = x$,　　　(b) $y = x + 1$,　　(c) $y = x + 2$,
   (d) $y = x + 3$,　　(e) $y = x - 1$,　　(f) $y = x - 2$,

   that lie in the square $-1 \leq x \leq 1$, $-1 \leq y \leq 1$. It is either impossible or extremely easy to do part (d); which? why?

3. For $-3 \leq x \leq 3$, sketch the graphs of

   (a) $y = (x + 1)\,\text{sgn}\, x$,　　　(b) $y = x\,\text{sgn}\,(x + 1)$.

4. For $-3 < x < 3$, sketch an (x, y) graph such that, for each x, the value of y is the slope of the graph in Fig. 3.6; for any x where there is no definite slope, use the average of the right-hand slope and the left-hand slope.

3.6 Graphs of Inequalities

Before we leave the visually instructive subject of graphs, let us investigate a few *inequalities* involving absolute values.

Consider, for example, the equation

$$|x| = 1$$

and the inequality

$$|x| \leq 1.$$

The equation has just two solutions, namely

$$x = 1 \quad \text{and} \quad x = -1;$$

but the inequality has the entire interval

$$-1 \leq x \leq 1$$

as solution.

Again, the equation

$$|x - 1| = 2$$

has just the two solutions

$$x = -1 \quad \text{and} \quad x = 3;$$

but the inequality

$$|x - 1| \leq 2$$

has, as solutions, each value x satisfying

$$-1 \leq x \leq 3.$$

See Fig. 3.12.

Figure 3.12. Graph of the inequality $|x - 1| \leq 2$

To find the solution of the inequality

$$(3.4) \qquad |x| + |y| \leq 1,$$

we consider values in each of the quadrants of the (x, y)-plane separately. Thus in the first quadrant, where

$$x \geq 0 \quad \text{and} \quad y \geq 0,$$

the inequality (3.4) is equivalent to

$$x + y \leq 1.$$

We draw the portion of the line

$$x + y = 1,$$

i.e., of the line

$$y = 1 - x,$$

that lies in the first quadrant; and since we seek the solution of

$$y \leq 1 - x,$$

our graph consists of those points in this quadrant that lie either on or below the line.

The graph is shown in Fig. 3.13, while the entire graph of the inequality (3.4) appears in Fig. 3.14.

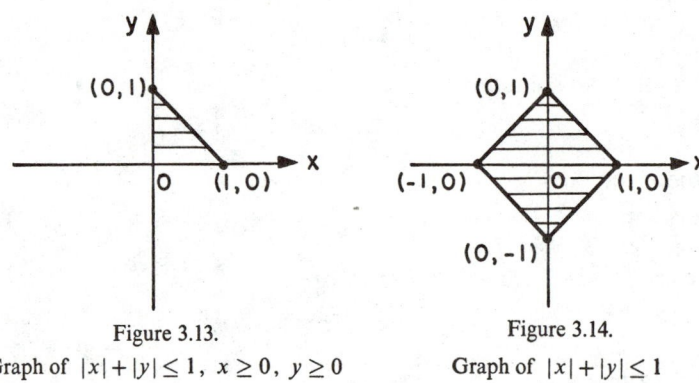

Figure 3.13.

Graph of $|x| + |y| \leq 1$, $x \geq 0$, $y \geq 0$

Figure 3.14.

Graph of $|x| + |y| \leq 1$

Figure 3.13 might be considered also as showing the graph of the solution of the set of inequalities

(3.5)
$$\begin{aligned} x + y &\leq 1 \\ x &\geq 0 \\ y &\geq 0. \end{aligned}$$

In Fig. 3.15 the shaded regions in (a), (b), and (c) comprise the solutions, respectively, of the first, second, and third inequality in (3.5); the triply shaded region in (d) is the simultaneous solution of all the inequalities (3.5). Thus, though the set of equations

$$\begin{aligned} x + y &= 1 \\ x &= 0 \\ y &= 0 \end{aligned}$$

is satisfied by no pair of values (x, y), the set of inequalities (3.5) has an entire *area* of solutions.

You will recall that, somewhat similarly, the equation $|x| = 1$ has only two solutions while the inequality $|x| \leq 1$ has an entire interval of solutions. Thus the set of all solutions of an inequality (or set of inequalities) is often much "richer" than the set of all solutions of the corresponding equation (or set of equations).

Exercises

1. In Fig. 3.15(d), the plane is separated into seven differently shaded portions. Each, together with its boundary, constitutes the solution of a set of three inequalities; for example, one set is $x \leq 0, y \geq 0, x + y \geq 1$.

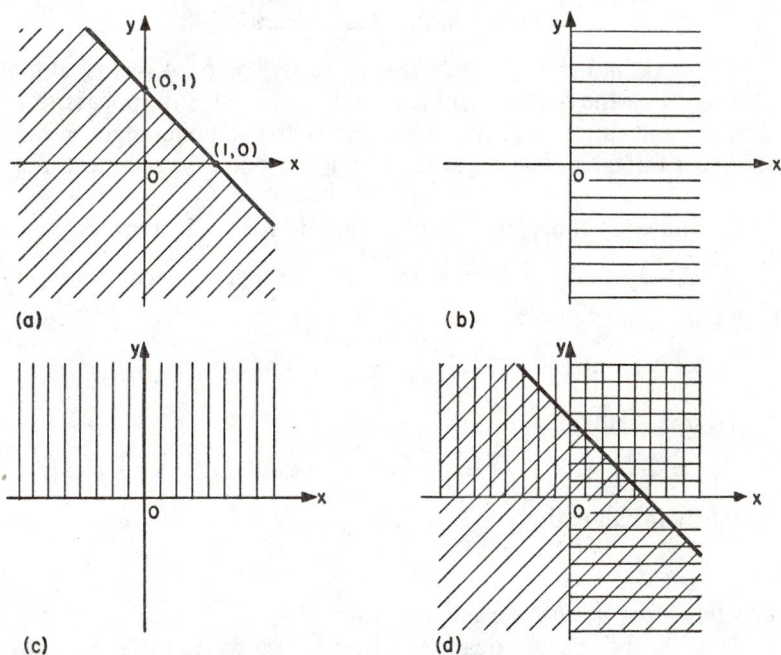

Figure 3.15. Incomplete graph of (a) $x + y \leq 1$, (b) $x \geq 0$, and (c) $y \geq 0$; triply shaded graph (d) of the intersection of (a), (b), and (c)

Give sets of inequalities for each of the areas.

2. Sketch the graphs of

    (a) $|x| - |y| \geq 1$ for $-2 \leq x \leq 2$,   (b) $|x| + 2|y| \leq 1$.

3. Sketch the graph of the system

$$y \leq x + 3$$
$$-2 \leq x \leq 2$$
$$y \geq 0.$$

4. Sketch the graph of the system

$$2x + y \leq 5$$
$$x - y \leq 1$$
$$x + 2y \leq 7.$$

5. Sketch the graph of the system

$$2x + y \geq 5$$
$$x - y \geq 1$$
$$x + 2y \geq 7.$$

3.7 Algebraic Characterization

Our next, and last, characterization of $|a|$ perhaps seems, at first glance, to be the *least* desirable as a definition, for it appears to be devious and unnatural; but it has the virtue of being algebraically the most tractable, and accordingly it is the one we shall most often use.

To this end, consider the following: If $a = -2$, then

$$a^2 = (-2)^2 = 4, \quad \sqrt{4} = 2 = |a|, \quad \text{so that} \quad \sqrt{a^2} = |a|;$$

if $a = 0$, then

$$a^2 = 0^2 = 0, \quad \sqrt{0} = 0 = |a|, \quad \text{so that} \quad \sqrt{a^2} = |a|;$$

and if $a = 2$, then

$$a^2 = 2^2 = 4, \quad \sqrt{4} = 2 = |a|, \quad \text{so that} \quad \sqrt{a^2} = |a|.$$

Similarly for all real a,

$$\sqrt{a^2} = |a|,$$

and this is our algebraic characterization of $|a|$.

Actually the characterization $|a| = \sqrt{a^2}$ expresses the special case $b = 0$ of the Pythagorean relationship,

$$c = \sqrt{a^2 + b^2},$$

between the lengths a, b of the sides and the length c of the hypotenuse of a right triangle. Thus the absolute value of the real number a can be interpreted as the *length, or magnitude, of the line segment from the origin to the point representing a on the number scale*

(Fig. 1.1). The reader who continues to study mathematics will soon learn that this last characterization of absolute value (and the accompanying geometric concept of length) can be appropriately generalized so that absolute value may be defined also for other mathematical objects, e.g., vectors and complex numbers.

Two points to note in the foregoing algebraic expression of $|a|$ are, first, that a^2 is nonnegative, so that its square roots are real, and secondly, that by definition the symbol $\sqrt{\ }$ means the nonnegative square root. To be sure, $(\pm 2)^2 = 4$, so that 4 has two square roots—namely ± 2; but in algebraic manipulations the symbol $\sqrt{4}$ denotes only 2, not -2. For example, consider the equations

$$5 + \sqrt{4} = 7, \quad 5 - \sqrt{4} = 3,$$

and

$$5 - \sqrt{4} = 7, \quad 5 + \sqrt{4} = 3.$$

The first two of these equations we consider to be valid, and the second two invalid, but only because the symbol $\sqrt{\ }$ always means the nonnegative square root. It is for this reason, also, that $\pm$ appears before the symbol $\sqrt{\ }$ in the familiar formula for the solutions of the quadratic equation

$$ax^2 + bx + c = 0, \quad a \neq 0,$$

namely

$$x = \frac{-b \pm \sqrt{b^2 - 4ac}}{2a}.$$

Exercises

1. You can see from the Pythagorean relationship that, for the equation
$$x^2 + y^2 = r^2,$$
the solution set (locus) of points (x, y) constitutes the circle with center at the origin and radius r. Determine the solution set of values (x, y) for the inequality
$$x^2 + y^2 \leq 25.$$

2. The absolute value of the complex number $x + iy$ is defined by
$$|x + iy| = \sqrt{x^2 + y^2}.$$
Representing $x + iy$ by the point having coordinates (x, y) in the plane, determine the solution set for
$$1 \leq |x + iy| \leq 2.$$

3. Determine the locus of points $x + iy$ for which
$$|x + iy + 1| = |x + iy - 1|.$$

3.8 The "Triangle" Inequality

The inequality (3.6) in Theorem 3.2 below, to which we have referred earlier in this chapter, is often called the "triangle" inequality for geometric reasons that will be discussed further in Chapter 4. Here is a complete statement concerning the inequality:

THEOREM 3.2. *For all real numbers a and b,*

$$(3.6) \qquad |a| + |b| \geq |a + b|.$$

The sign of equality holds if and only if $ab \geq 0$, that is, if and only if a and b are either both ≥ 0 or both ≤ 0.

For example, if $a = 5$ and $b = -2$, then

$$|a + b| = |5 + (-2)| = 3,$$

while

$$|a| + |b| = |5| + |-2| = 7.$$

But if $a = -5$ and $b = -2$, then

$$|a + b| = |(-5) + (-2)| = 7$$

and

$$|a| + |b| = |-5| + |-2| = 7.$$

You can grasp the validity of the inequality (3.6) intuitively by noting that if a and b are of opposite sign, then on the right-hand side of the inequality the numbers a and b will "work against each other" and emerge from the encounter with diminished joint magnitude, whereas on the left-hand side they are forced to "pull together" positively from the start.

On the same basis, you can understand the inequality

$$(3.7) \qquad |a - b| \geq \Big| |a| - |b| \Big|,$$

in which the sign of equality holds again if and only if $ab \geq 0$ (i.e., a and b are both ≥ 0 or both ≤ 0). Here, on the right-hand side of the inequality, the numbers a and b are always forced to "settle their difference," whereas on the left-hand side, if a and b are of opposite sign, then they will pull together to increase their joint magnitude.

Inequalities (3.6) and (3.7) can be proved by using the algebraic definition $|a| = \sqrt{a^2}$. Thus inequality (3.6) can be restated equivalently as

$$(3.8) \qquad \sqrt{a^2} + \sqrt{b^2} \geq \sqrt{(a + b)^2}.$$

Now, (3.8) is equivalent to

$$(3.9) \qquad (\sqrt{a^2} + \sqrt{b^2})^2 \geq (a + b)^2 ;$$

that is, the validity of either (3.8) or (3.9) implies that of the other. Thus (3.9) follows from (3.8) by squaring (see Theorem 2.5), and (3.8) follows from (3.9) by taking nonnegative square roots (see Theorem 2.7).

Next, (3.9) can be rewritten as

$$a^2 + 2\sqrt{a^2 b^2} + b^2 \geq a^2 + 2ab + b^2 .$$

Accordingly, (3.9) is equivalent to

$$(3.10) \qquad \sqrt{a^2 b^2} \geq ab$$

by the inequality rules for addition, subtraction, and multiplication by a positive number. But

$$\sqrt{a^2 b^2} = \sqrt{(ab)^2} = |ab| ,$$

so that (3.10) is equivalent to

$$(3.11) \qquad |ab| \geq ab.$$

Thus (3.6) is equivalent to (3.11).

Now, (3.11) is valid by Theorem 3.1, which states that any real number is less than or equal to its absolute value. The sign of equality in (3.11) holds if and only if $ab \geq 0$. Therefore the inequality (3.6), which is equivalent to (3.11), is also valid, and the sign of equality holds if and only if $ab \geq 0$.

The inequality (3.7) can be proved similarly after it has been written in the equivalent form

$$\sqrt{(a - b)^2} \geq \sqrt{(\sqrt{a^2} - \sqrt{b^2})^2} .$$

However, it is interesting to note that (3.7) can also be derived directly from (3.6). Thus, substitution of the real number $a - b$ for the arbitrary real number a in (3.6) yields

$$|a - b| + |b| \geq |a - b + b| ,$$

or

$$|a - b| + |b| \geq |a| ,$$

whence

$$(3.12) \qquad |a - b| \geq |a| - |b|$$

by Theorem 2.4, the subtraction rule. Similarly, substitution of $b - a$

for b in (3.6) yields

$$|a| + |b - a| \geq |a + b - a|,$$

or

$$|a| + |a - b| \geq |b|,$$

whence

(3.13) $$|a - b| \geq |b| - |a|.$$

Since

$$\begin{aligned}\left||a| - |b|\right| &= \max\{(|a| - |b|), -(|a| - |b|)\} \\ &= \max\{(|a| - |b|), (|b| - |a|)\},\end{aligned}$$

it follows from (3.12) and (3.13) that

$$|a - b| \geq \left||a| - |b|\right|.$$

Hence (3.7) is valid, the sign of equality holding if and only if it holds either in (3.12) or in (3.13).

Since $a - b$ was substituted for a in deriving (3.12), the sign of equality holds in (3.12) if and only if $(a - b)b \geq 0$, i.e., $ab \geq b^2$. The latter inequality is valid if and only if $ab \geq 0$ and $|a| \geq |b|$. Similarly, the sign of equality holds in (3.13) if and only if $a(b - a) \geq 0$, that is, $ab \geq a^2$. This is valid if and only if $ab \geq 0$ and $|b| \geq |a|$. Since at least one of the inequalities $|a| \geq |b|$ and $|b| \geq |a|$ holds, it follows that the sign of equality holds in (3.7) if and only if $ab \geq 0$.

It might be noted that since b represents an arbitrary real number—positive, zero, or negative—the inequalities (3.6) and (3.7) still hold if b is replaced by $-b$. We thus obtain

(3.14) $$|a| + |b| \geq |a - b|$$

from (3.6), and

(3.15) $$|a + b| \geq \left||a| - |b|\right|$$

from (3.7), with the signs of equality holding if and only if $a(-b) \geq 0$, that is, $ab \leq 0$. Together, inequalities (3.6), (3.7), (3.14), and (3.15) can be written as

(3.16) $$|a| + |b| \geq |a \pm b| \geq \left||a| - |b|\right|.$$

Exercises

1. From the algebraic characterization of absolute value, show for all real a, b that

$$\text{(a)} \quad |-a| = |a|, \qquad \text{(b)} \quad |ab| = |a| \cdot |b|,$$

and if $b \neq 0$ that

$$\text{(c)} \quad \left| \frac{a}{b} \right| = \frac{|a|}{|b|}.$$

2. Determine whether it is the sign $<$ or the sign $=$ that holds in the mixed inequality $|a - b| \geq \big| |a| - |b| \big|$, if

  (a) $a = \pi$, $b = 2\pi$;   (d) $a = -9$, $b = -10$;
  (b) $a = -\pi$, $b = \sqrt{2}$;   (e) $a = 9$, $b = -10$.
  (c) $a = 2$, $b = 0$;

3. Determine whether it is the sign $>$ or the sign $=$ that holds in the mixed inequality $|a| + |b| \geq |a + b|$, if

  (a) $a = 3$, $b = -2$;   (d) $a = 0$, $b = -2$;
  (b) $a = -3$, $b = -2$;   (e) $a = 0$, $b = 0$.
  (c) $a = 3$, $b = 2$;

4. Repeat Exercise 2 for the inequality $|a + b| \geq \big| |a| - |b| \big|$.

5. Repeat Exercise 3 for the inequality $|a| + |b| \geq |a - b|$.

6. Show that the inequality $|a - b| \geq \big| |a| - |b| \big|$ is equivalent to the inequality $|ab| \geq ab$.

7. Show that if $ab \geq 0$ then $ab \geq \min \{a^2, b^2\}$.

8. Show that each of the other characteristic properties of $|a|$, given in this chapter, follows from $|a| = \sqrt{a^2}$.

CHAPTER FOUR

The Classical Inequalities

4.1 Introduction

Now that we have forged our basic tools, we shall demonstrate more of the magic of mathematics. As an artist evokes, out of a few lines on a canvas, scenes of great beauty, and as a musician conjures up exquisite melodies from combinations of a few notes, so the mathematician with a few penetrating logical steps portrays results of simple elegance. Often, like the product of the magician's wand, these results seem quite mysterious, despite their simplicity, until their origin is perceived.

In this chapter, we shall employ the basic results derived in the previous chapters to obtain some of the most famous inequalities in the field of mathematical analysis. These inequalities are the everyday working tools of the specialist in this branch of mathematics.

In Chapter 5, we shall then show how these new relationships may be used to solve a number of interesting problems that, at first sight, seem far removed from algebra and inequalities. The applications are continued in Chapter 6, where we discuss and extend the notion of distance.

This, indeed, is one of the fascinations of mathematics—that simple ideas applied one after the other, in the proper order, yield results that never could have been envisaged at the outset.

4.2 The Inequality of the Arithmetic and Geometric Means

(a) *Mathematical Experimentation.* Given two nonnegative numbers, say 1 and 2, let us obtain their "mean" in the following two ways: the *arithmetic mean* (or *half their sum*), usually called "average,"

$$\frac{1 + 2}{2} = 1.5,$$

and the *geometric mean* (or *square root of their product*)

$$\sqrt{1 \cdot 2} = 1.41 \cdots.$$

Observe that $1.5 > 1.41 \cdots$. Similarly, if we start with the numbers 3 and 9, for the arithmetic mean we obtain the value $\frac{1}{2}(3 + 9) = 6$, and for the geometric mean we get the value $\sqrt{27} = 5.19 \cdots$. Note that $6 > 5.19 \cdots$. Continuing with various pairs of nonnegative numbers chosen at random, say 11 and 13, $\frac{1}{2}$ and $\frac{1}{4}$, and so on, we observe in each case that the arithmetic mean is greater than the geometric mean.

Can we safely generalize this discovery and reach some conclusions? Our mathematical nose begins to twitch as we scent a theorem. Maybe the result is true for all pairs of nonnegative numbers! In other words, we conjecture that the arithmetic mean of two nonnegative numbers is always at least as great as their geometric mean. We shall express this conjecture in terms of algebraic symbols and we shall see in subsection (b) that our conjecture is true. We may therefore state it as

THEOREM 4.1. *For any nonnegative numbers a and b,*

(4.1) $$\frac{a + b}{2} \geq \sqrt{ab}.$$

The sign of equality holds if and only if $a = b$.

Note that if one of the two numbers were positive and the other negative, then (4.1) would be meaningless since its right-hand side would be imaginary.† If both numbers were negative, then the left-hand side of (4.1) would be negative and the right-hand side positive, so that the theorem would not be valid.

---

† The notion of inequality is not directly applicable to imaginary numbers, but only to their absolute values.

The type of experimentation that led us to Theorem 4.1 represents the sort of trial-and-error methods often used by mathematicians on the trail of theorems. Formerly, it was quite laborious work. Nowadays, with the modern digital computer to aid mathematical experimentation, we can test thousands and millions of cases in a few hours. In this way, we obtain valuable clues to general mathematical truths.

Exercises

1. Determine the geometric mean and the arithmetic mean for the following pairs of numbers:

    (a) 2, 8,   (b) 3, 12,   (c) 4, 9,   (d) 0, 20.

2. If p is nonnegative, determine the geometric mean and the arithmetic mean of the following pairs of numbers:

    (a) $p, 9p$,   (b) $0, p$,   (c) $2, 2p^2$.

(b) *Proof of the Arithmetic-mean–Geometric-mean Inequality for Two Numbers.* Since square roots are a bit bothersome, let us eliminate them by writing

$$(4.2) \qquad a = c^2, \qquad b = d^2.$$

This is permissible because a and b were assumed, in Theorem 4.1, to be nonnegative. The relationship (4.1) that we wish to prove for arbitrary *nonnegative a, b* then becomes

$$(4.3) \qquad \frac{c^2 + d^2}{2} \geq cd$$

for arbitrary *real c, d.* Now (4.3) is true if and only if

$$(4.4) \qquad \frac{c^2 + d^2}{2} - cd \geq 0,$$

which is equivalent to

$$(4.5) \qquad c^2 + d^2 - 2cd \geq 0$$

as a result of our elementary rules for dealing with inequalities.

We now recognize a familiar friend, namely

$$(4.6) \qquad c^2 + d^2 - 2cd = (c - d)^2,$$

so that (4.5) is equivalent to

$$(4.7) \qquad (c - d)^2 \geq 0.$$

Since, by Theorem 1.3, the square of any real number is nonnegative, we see that (4.7) is indeed true. Thus (4.5) is a valid inequality, and hence (4.4), (4.3), and (4.1) are also valid. The sign of equality holds in (4.7), and therefore in (4.1), if and only if $c - d = 0$, that is, $c = d$, or, equivalently, if and only if $a = b$.

Note that, while the inequality (4.1) of Theorem 4.1 applies only to nonnegative numbers a, b, the foregoing proof shows that inequality (4.3) is valid for all real numbers c, d, the sign of equality holding if and only if $c = d$. You will observe that the results of Secs. 4.4 and 6.6 also are valid for all real numbers, not merely nonnegative numbers; this fact increases the geometric significance of those results.

(c) *A Geometric Proof.* Let us now show that Theorem 4.1 can also be obtained geometrically by means of a simple comparison of areas. Consider the graph of $y = x$, as shown in Fig. 4.1. Let S and T be points on the line $y = x$, with coordinates (c, c) and (d, d), respectively, and consider the points $P:(c, 0)$, $Q:(0, d)$, and $R:(c, d)$, as shown. Since OP is of length c, PS has the same length c. Then the area of the triangle OPS is $c^2/2$, i.e., one half of the base times the altitude. Similarly, the area of the right triangle OQT is $d^2/2$.

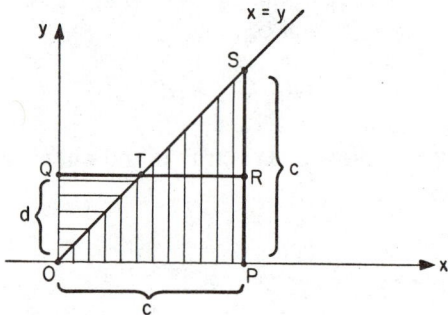

Figure 4.1. Geometric proof of the inequality $\dfrac{c^2 + d^2}{2} \geq cd$

Now examine the rectangle $OPRQ$. Its area is completely covered by the triangles OPS and OQT, so that

(4.8)   area (OPS) + area $(OQT) \geq$ area $(OPRQ)$.

Since the area of $OPRQ$ is cd, length times width, we may write (4.8)

in algebraic symbols as

(4.9)
$$\frac{c^2 + d^2}{2} \geq cd.$$

Now the inequality (4.9) is identical with the inequality (4.3), so that our geometric proof is complete.

Furthermore, we see that there is equality only when the triangle *TRS* has area zero, that is, only when S and T coincide, so that $c = d$.

(d) *A Geometric Generalization.* A little thought will show that the foregoing arguments remain valid even in cases where the curve *OTS* is not a straight line. Consider the diagram shown in Fig. 4.2. It is still true that

(4.10)   area (OPS) + area $(OQT) \geq$ area $(OPRQ)$.

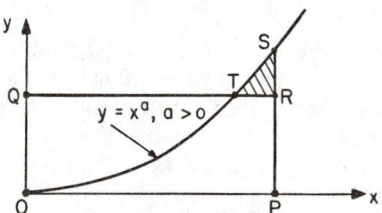

Figure 4.2. A more general geometric inequality

When you have studied calculus and have learned how to evaluate the area underneath simple curves, such as $y = x^a$, for arbitrary positive a, you will find that this process will yield a number of interesting inequalities in a very simple fashion. In later sections of this chapter, we shall obtain some of these inequalities in a different way.

Exercises

1. Let a and b be the lengths of a pair of adjacent segments on a line, and draw a semicircle with the two segments together as its diameter, as in

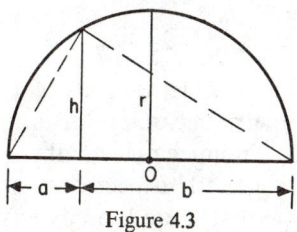

Figure 4.3

Fig. 4.3. Show that the radius r of the circle is the arithmetic mean of a and b, and that the perpendicular distance h is their geometric mean.

2. An average that occurs quite naturally in optics and in the study of electrical networks is the *harmonic mean*. For two given positive quantities a and b, the quantity c determined by means of the relationship

$$\frac{2}{c} = \frac{1}{a} + \frac{1}{b}$$

is called the harmonic mean. Solving this equation for c, we obtain

$$c = \frac{2}{1/a + 1/b} = \frac{2ab}{a + b}.$$

Show that the harmonic mean is less than or equal to the arithmetic mean, and also less than or equal to the geometric mean, with equality if and only if $a = b$; i.e., show that

$$\frac{a + b}{2} \geq \sqrt{ab} \geq \frac{2ab}{a + b}.$$

3. Determine the harmonic, geometric, and arithmetic means of the pairs

(a) 2, 8;   (b) 3, 12;   (c) 4, 9;   (d) 5, 7;   (e) 6, 6.

4. The relationship between distance d, rate r, and time t is $d = rt$. Show that if in traveling from one town to another you go half the distance at rate r_1, and half at rate r_2, then your average rate is the harmonic mean of r_1 and r_2, but that if you travel half the time at rate r_1 and half the time at rate r_2, then your average rate is their arithmetic mean. If $r_1 \neq r_2$, which method would get you there sooner?

5. Use the result of Theorem 4.1 to solve Exercise 2 on page 22.

(e) *The Arithmetic-mean–Geometric-mean Inequality for Three Numbers.* Let us now perform some further experimentation. Taking three nonnegative numbers, say 1, 2, and 4, let us compute their arithmetic mean—the simple average—as before:

$$\frac{1 + 2 + 4}{3} = 2.33\cdots.$$

Let us also compute their geometric mean, i.e., the cube root of their product:

$$\sqrt[3]{1 \cdot 2 \cdot 4} = 2.$$

We observe that the arithmetic mean of these three numbers is greater than their geometric mean. Performing a number of such experiments with triplets of nonnegative numbers, we constantly observe the same result. We begin to suspect that we have found another theorem. Can it be true that there is an extension of Theorem 4.1, a result asserting that the arithmetic mean of three nonnegative quantities is at least as great as their geometric mean?

We wish to prove

THEOREM 4.2. *For any three nonnegative numbers a, b, c,*

$$(4.11) \qquad \frac{a+b+c}{3} \geq \sqrt[3]{abc}.$$

The sign of equality holds if and only if $a = b = c$.

To remove the cube roots, let us set

$$(4.12) \qquad a = x^3, \qquad b = y^3, \qquad c = z^3.$$

Substituting these values for a, b, and c in (4.11), we obtain

$$(4.13) \qquad \frac{x^3 + y^3 + z^3}{3} \geq xyz,$$

which is equivalent to

$$(4.14) \qquad x^3 + y^3 + z^3 - 3xyz \geq 0.$$

We shall prove Theorem 4.2 by proving the validity of (4.14) for arbitrary nonnegative x, y, z.

Once again we have an expression that can be factored. Its factorization is not as common as the one used before, but is still a quite useful one. We assert that

$$(4.15) \qquad \begin{aligned} x^3 + y^3 + z^3 - 3xyz \\ = (x+y+z)(x^2 + y^2 + z^2 - xy - xz - yz), \end{aligned}$$

a result that can be verified by multiplication.

Since $x + y + z$ is nonnegative, the first factor on the right in (4.15) is positive unless $x = y = z = 0$. In order to demonstrate (4.14), it is sufficient to show that the second factor also is nonnegative, i.e., that

$$(4.16) \qquad x^2 + y^2 + z^2 - xy - xz - yz \geq 0.$$

Referring back to the inequality $(x - y)^2 = x^2 + y^2 - 2xy \geq 0$ already used in the algebraic proof of the arithmetic-mean–geometric-mean inequality for two numbers [see subsection (b)], we see that the inequality (4.16) can be derived from this in the following fashion. Write

$$(4.17) \quad x^2 + y^2 \geq 2xy, \qquad x^2 + z^2 \geq 2xz, \qquad y^2 + z^2 \geq 2yz,$$

and add the three inequalities. The result is

$$(4.18) \qquad 2(x^2 + y^2 + z^2) \geq 2(xy + xz + yz),$$

which is equivalent to the desired inequality (4.16). The sign of equality holds if and only if $x = y = z$.

Since (4.16) is a valid inequality, and since $x + y + z \geq 0$, it follows that the left-hand side of (4.15) is also ≥ 0; that is, the inequality (4.14) is valid. Since (4.14) is equivalent to (4.11), we have now proved that the arithmetic-mean–geometric-mean inequality is valid for means of three numbers; the condition $x = y = z$ under which the sign of equality holds in (4.14), and therefore in (4.11), is equivalent to the condition $a = b = c$.

(f) *The Arithmetic-mean–Geometric-mean Inequality for n Numbers.* Emboldened by this success, let us conjecture that the results we have obtained for means of two and three numbers are merely special cases of a general theorem valid for any number of positive quantities. If this conjecture is true, then we have the following result:

THEOREM 4.3. *For any n nonnegative numbers* $a_1, a_2, \ldots, a_n$,

$$(4.19) \qquad \frac{a_1 + a_2 + \cdots + a_n}{n} \geq \sqrt[n]{a_1 a_2 \cdots a_n}.$$

The sign of equality holds if and only if $a_1 = a_2 = \cdots = a_n$.

This is the famous inequality connecting the arithmetic mean of n quantities with the geometric mean of the n quantities, and is, indeed, true. We have concentrated on this inequality for several reasons. In the first place, it is a fascinating one, and one that can be established in a large number of interesting ways; there are literally dozens of different proofs based on ideas from a great variety of sources. In the second place, it can be used as the fundamental theorem of the theory of inequalities, the keystone on which many other very important results rest. In the third place, as you will see in Chapter 5, we can use some of its consequences to solve a number of maximization and minimization problems.

In attempting to prove the general inequality, a first thought may be to continue along the preceding lines, using another algebraic factorization for $n = 4$, still another one for $n = 5$, and so on. But this approach is neither attractive nor feasible. As a matter of fact, no simple proof along these lines exists.

Instead, we shall present a simple proof based on two applications of mathematical induction; one, a "forward" induction, will lead to the desired result for all the integers n that are powers of two, i.e., for $n = 2^k$; the other, a "backward" induction (from any positive integer to the preceding one), together with the forward induction will enable us to establish the result for all positive integers n.

(i) *Forward Induction.* The method we shall now employ will illustrate an amusing variant of the fundamental technique of proof by mathematical induction, which we have previously discussed in Sec. 2.6 (Chapter 2).

Let us start with the result for $n = 2$, namely

$$(4.20) \qquad \frac{a+b}{2} \geq \sqrt{ab},$$

which, by Theorem 4.1, is valid for all nonnegative a and b, and use some mathematical ingenuity. Although there are very many simple proofs of Theorem 4.3, they all possess the common ingredient of ingenuity.

Set

$$a = \frac{a_1 + a_2}{2}, \qquad b = \frac{a_3 + a_4}{2},$$

where a_1, a_2, a_3, a_4 are nonnegative numbers. Substituting these values for a and b in (4.20), we obtain the inequality

$$\frac{\dfrac{a_1+a_2}{2} + \dfrac{a_3+a_4}{2}}{2} \geq \sqrt{\left(\frac{a_1+a_2}{2}\right)\left(\frac{a_3+a_4}{2}\right)},$$

or

$$(4.21) \qquad \frac{a_1+a_2+a_3+a_4}{4} \geq \sqrt{\left(\frac{a_1+a_2}{2}\right)\left(\frac{a_3+a_4}{2}\right)}.$$

Since the left-hand side of the inequality (4.21) has the desired form (see Theorem 4.3), let us concentrate on the right-hand side. Using the valid inequalities

$$(4.22) \qquad \frac{a_1+a_2}{2} \geq \sqrt{a_1 a_2}, \qquad \frac{a_3+a_4}{2} \geq \sqrt{a_3 a_4},$$

and the transitivity rule (Theorem 2.1), we obtain from (4.21) the further inequality

$$(4.23) \qquad \frac{a_1 + a_2 + a_3 + a_4}{4} \geq \sqrt{\sqrt{a_1 a_2} \sqrt{a_3 a_4}} \geq (a_1 a_2 a_3 a_4)^{1/4}.$$

But this is precisely the desired result for four nonnegative quantities! The arithmetic mean is greater than or equal to the geometric mean for $n = 4$.

The sign of equality holds in (4.21) if and only if

$$\frac{a_1 + a_2}{2} = \frac{a_3 + a_4}{2},$$

and in (4.22) if and only if $a_1 = a_2$, $a_3 = a_4$; consequently, it holds in (4.23) if and only if $a_1 = a_2 = a_3 = a_4$.

Nothing stops us from repeating the foregoing trick. Set

$$a_1 = \frac{b_1 + b_2}{2}, \qquad a_2 = \frac{b_3 + b_4}{2}, \qquad a_3 = \frac{b_5 + b_6}{2}, \qquad a_4 = \frac{b_7 + b_8}{2},$$

where the b_i $(i = 1, 2, \ldots, 8)$ are nonnegative quantities. Substituting in (4.23), we have

$$\frac{b_1 + b_2 + \cdots + b_8}{8} \geq \left[\left(\frac{b_1 + b_2}{2} \right) \frac{b_3 + b_4}{2} \left(\frac{b_5 + b_6}{2} \right) \left(\frac{b_7 + b_8}{2} \right) \right]^{1/4}.$$

Using the inequalities

$$\frac{b_1 + b_2}{2} \geq \sqrt{b_1 b_2}, \qquad \ldots, \qquad \frac{b_7 + b_8}{2} \geq \sqrt{b_7 b_8},$$

and the transitivity rule, we obtain

$$\frac{b_1 + b_2 + \cdots + b_8}{8} \geq (\sqrt{b_1 b_2} \sqrt{b_3 b_4} \sqrt{b_5 b_6} \sqrt{b_7 b_8})^{1/4}$$

$$\geq (b_1 b_2 \cdots b_8)^{1/8},$$

the desired result for eight quantities. The sign of equality holds if and only if all the b_i are equal.

Continuing in this way, we clearly can establish the inequality for all values of n that are powers of two, i.e., for $n = 2, 4, 8, 16, \ldots$. To tie the result down rigorously, we use *mathematical induction*. The main step consists of proving the following result:

The arithmetic-mean–geometric-mean inequality is valid for all n of the form 2^k, $k = 1, 2, \ldots$.

PROOF. We already know that the result is true for $n = 2 = 2^1$, i.e., for $k = 1$, and indeed for $n = 2^2$ and 2^3. Let us assume that the result is true for an integer n of the form 2^k, and then establish it for 2^{k+1}. Since $2^{k+1} = 2 \cdot 2^k$, this means that we shall prove that the result is true for $2n$.

Thus we have assumed that

$$(4.24) \qquad \frac{a_1 + a_2 + \cdots + a_n}{n} \geq (a_1 a_2 \cdots a_n)^{1/n}$$

for any set of nonnegative quantities $a_1, a_2, \ldots, a_n$, where $n = 2^k$. Choose $a_i \ (i = 1, 2, \ldots, n)$ to have the following values:

$$a_1 = \frac{b_1 + b_2}{2}, \qquad a_2 = \frac{b_3 + b_4}{2}, \qquad \ldots, \qquad a_n = \frac{b_{2n-1} + b_{2n}}{2},$$

where the $2n$ numbers $b_j \ (j = 1, 2, \ldots, 2n)$ are given nonnegative numbers, and substitute in (4.24). Proceeding as before, we finally obtain

$$\frac{b_1 + b_2 + \cdots + b_{2n}}{2n} \geq (b_1 b_2 \cdots b_{2n})^{1/2n}.$$

As before, the sign of equality holds if and only if all the b_i are equal. Hence we have established the desired result for $2n$, or 2^{k+1}.

Thus, since the inequality is valid for $k = 1$, the principle of (forward) mathematical induction asserts that the result is true for all positive integers k, and hence that the inequality (4.24) holds for all n that are powers of two.

(ii) *Backward Induction.* Now that we have established the result for those integers that are powers of two, how do we establish it for the full set of positive integers?

Another procedure is required. Consider the case $n = 3$, for which, of course, we have already established the result by a different method. Using the relationship for $n = 2^2 = 4$,

$$(4.25) \qquad \frac{a_1 + a_2 + a_3 + a_4}{4} \geq (a_1 a_2 a_3 a_4)^{1/4},$$

which has already been established by forward induction, let us see if we can derive the corresponding result for $n = 3$.

We accomplish this by means of the important technique of *specialization*. Starting with (4.25), we choose the quantities a_1, a_2, a_3, and a_4 in a special way. Set

(4.26) $a_1 = b_1,$ $a_2 = b_2,$ $a_3 = b_3,$

and ask for the value a_4 that yields the equality

$$\frac{a_1 + a_2 + a_3 + a_4}{4} = \frac{b_1 + b_2 + b_3}{3}.$$

By the values given in (4.26), this requires that

$$\frac{b_1 + b_2 + b_3 + a_4}{4} = \frac{b_1 + b_2 + b_3}{3},$$

whence

$$a_4 = \frac{4}{3}(b_1 + b_2 + b_3) - (b_1 + b_2 + b_3) = \frac{b_1 + b_2 + b_3}{3}.$$

Substituting these particular values for the a_i in (4.25), we derive the relationship

$$\frac{b_1 + b_2 + b_3}{3} \geq \sqrt[4]{b_1 b_2 b_3 \left(\frac{b_1 + b_2 + b_3}{3}\right)}.$$

Raising both sides to the fourth power, we obtain

$$\left(\frac{b_1 + b_2 + b_3}{3}\right)^4 \geq b_1 b_2 b_3 \left(\frac{b_1 + b_2 + b_3}{3}\right),$$

or finally, dividing by $(b_1 + b_2 + b_3)/3$,

$$\left(\frac{b_1 + b_2 + b_3}{3}\right)^3 \geq b_1 b_2 b_3,$$

which is equivalent to the desired result

(4.27) $\dfrac{b_1 + b_2 + b_3}{3} \geq \sqrt[3]{b_1 b_2 b_3}.$

Since equality holds in (4.25) if and only if $a_1 = a_2 = a_3 = a_4$, it follows that equality holds in (4.27) if and only if $b_1 = b_2 = b_3$.

In order to extend this method to the general case, we shall employ an inductive technique, but an inductive technique of nonstandard type. Instead of proving that if the result holds for n then it holds for $n + 1$, we shall prove that if it holds for n then it holds for $n - 1$. Since we already know that it holds for $n = 2^k$ $(k = 1, 2, \ldots)$, this method will yield the theorem in all generality.

Let us now show that if the result holds for n, then it holds for $n - 1$. To do this, we repeat the trick of specialization that we used above. Let

$$(4.28) \qquad a_1 = b_1, \qquad a_2 = b_2, \qquad \ldots, \qquad a_{n-1} = b_{n-1},$$

and determine a_n by the requirement that

$$\frac{a_1 + a_2 + \cdots + a_n}{n} = \frac{b_1 + b_2 + \cdots + b_{n-1}}{n - 1};$$

using the values (4.28) and solving for a_n, we get

$$(4.29) \qquad a_n = \frac{b_1 + b_2 + \cdots + b_{n-1}}{n - 1}.$$

We have assumed that the inequality

$$\frac{a_1 + a_2 + \cdots + a_n}{n} \geq \sqrt[n]{a_1 a_2 \cdots a_n}$$

for the n nonnegative quantities $a_1, a_2, \ldots, a_n$ is valid; substituting the values (4.28) and (4.29) for the a_i, we have

$$\frac{b_1 + b_2 + \cdots + b_{n-1}}{n - 1} \geq \sqrt[n]{b_1 b_2 \cdots b_{n-1} \left(\frac{b_1 + b_2 + \cdots + b_{n-1}}{n - 1} \right)}$$

Raising both sides to the nth power and simplifying, we obtain the inequality

$$\left(\frac{b_1 + b_2 + \cdots + b_{n-1}}{n - 1} \right)^{n-1} \geq b_1 b_2 \cdots b_{n-1},$$

which is equivalent to the desired result

$$\frac{b_1 + b_2 + \cdots + b_{n-1}}{n - 1} \geq \sqrt[n-1]{b_1 b_2 \cdots b_{n-1}}.$$

As before, equality holds if and only if $b_1 = b_2 = \cdots = b_{n-1}$, and the proof of Theorem 4.3 is complete.

4.3 Generalizations of the Arithmetic-mean-Geometric-mean Inequality

We shall now show that a number of results that appear to be generalizations of the fundamental arithmetic-mean–geometric-mean theorem, derived above, are actually special cases.

First, let us take the arithmetic-mean–geometric-mean inequality

$$\frac{x_1 + x_2 + \cdots + x_n}{n} \geq (x_1 x_2 \cdots x_n)^{1/n}$$

and set the first m of the numbers x_i equal to the same nonnegative value x, and the remaining $n - m$ equal to a common nonnegative value y; that is,

$$x_1 = x_2 = \cdots = x_m = x, \qquad x_{m+1} = x_{m+2} = \cdots = x_n = y.$$

The arithmetic-mean–geometric-mean inequality for $x_1, x_2, \ldots, x_n$ becomes

$$\frac{mx + (n - m)y}{n} \geq (x^m y^{n-m})^{1/n}$$

or

$$\frac{mx}{n} + \left(1 - \frac{m}{n}\right)y \geq x^{m/n} y^{1-m/n}.$$

Here n is any positive integer, and m is any integer in the range $1 \leq m \leq n - 1$. It follows that m/n can represent any rational fraction r occurring in the interval $0 < r < 1$. Let us then write the foregoing inequality in the form

$$(4.30) \qquad rx + (1 - r)y \geq x^r y^{1-r},$$

a most important result for our subsequent purposes.

This inequality (4.30) is valid for any two nonnegative quantities x and y, and for any fraction r between 0 and 1. Equality occurs if and only if $x = y$.

Let r be denoted by $1/p$. Since $0 < r < 1$, we see that $p > 1$. Then

$$1 - r = 1 - \frac{1}{p} = \frac{p - 1}{p}.$$

Let q denote the quantity $p/(p - 1)$, so that $1/q = 1 - r$ and

$$\frac{1}{p} + \frac{1}{q} = 1.$$

The inequality (4.30) then has the form

(4.31)
$$\frac{x}{p} + \frac{y}{q} \geq x^{1/p} y^{1/q}.$$

To eliminate the fractional powers, set

(4.32)
$$x = a^p, \qquad y = b^q.$$

Then (4.31) assumes the form

(4.33)
$$\frac{a^p}{p} + \frac{b^q}{q} \geq ab,$$

where a and b are nonnegative numbers, and p and q are rational numbers satisfying $1/p + 1/q = 1$. There is equality if and only if

(4.34)
$$a^p = b^q.$$

Once we have defined what we mean by irrational numbers and by functions of the form x^r, where r is irrational, we can show, either directly or by means of a limiting procedure starting from the inequality (4.30), that the inequality (4.30) is actually valid for all r between 0 and 1, and therefore that (4.33) is valid for all $p > 1$, $q > 1$ satisfying $1/p + 1/q = 1$. If you wish to pursue this refinement further, you should first read *Numbers: Rational and Irrational* by Ivan Niven, to which we have already referred in Chapter 1.

Exercises

1. Show that if the values $y_1, y_2, \ldots, y_k$ are all nonnegative, and if $m_1, m_2, \ldots, m_k$ are positive integers, then

$$\frac{m_1 y_1 + m_2 y_2 + \cdots + m_k y_k}{m_1 + m_2 + \cdots + m_k} \geq (y_1^{m_1} y_2^{m_2} \cdots y_k^{m_k})^{1/(m_1 + m_2 + \cdots + m_k)}.$$

Show consequently that if $r_1, r_2, \ldots, r_k$ are proper fractions satisfying

$$r_1 + r_2 + \cdots + r_k = 1,$$

then

$$r_1 y_1 + r_2 y_2 + \cdots + r_k y_k \geq y_1^{r_1} y_2^{r_2} \ldots y_k^{r_k}.$$

2. An important average in statistics is the *root-mean-square*. For two nonnegative numbers a and b, the root-mean-square is the value

$$s = \sqrt{\frac{a^2 + b^2}{2}}.$$

For the pairs $(5, 12)$, $(0, 1)$, and (p, p), compute the arithmetic mean and the root-mean-square.

3. Show that the arithmetic mean of two positive numbers is less than or equal to their root-mean-square:

$$\frac{a+b}{2} \leq \sqrt{\frac{a^2 + b^2}{2}}.$$

Under what circumstance does the sign of equality hold? How does the root-mean-square compare with the geometric mean and with the harmonic mean?

4. Let $ABDC$ be a trapezoid with $AB = a$, $CD = b$ (see Fig. 4.4). Let O be the point of intersection of its diagonals. Show that

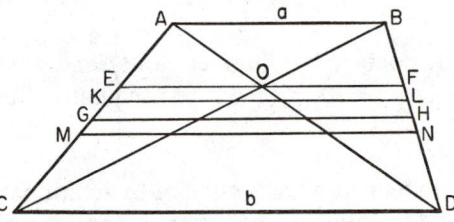

Figure 4.4. Geometric illustration of $\dfrac{2ab}{a+b} \leq \sqrt{ab} \leq \dfrac{a+b}{2} \leq \sqrt{\dfrac{a^2 + b^2}{2}}$

(a) The arithmetic mean $(a + b)/2$ of a and b is represented by the line segment GH parallel to the bases and halfway between them.

(b) The geometric mean $\sqrt{ab}$ is represented by the line segment KL parallel to the bases and situated so that trapezoids $ABLK$ and $KLDC$ are similar.

(c) The harmonic mean is represented by the line segment EF parallel to the bases and passing through O.

(d) The root-mean-square is represented by the line segment MN parallel to the bases and dividing the trapezoid $ABDC$ into two trapezoids of equal area.

4.4 The Cauchy Inequality

(a) *The Two-dimensional Version:* $(a^2 + b^2)(c^2 + d^2) \geq (ac + bd)^2$. Let us now introduce a new theme. As in a musical composition, this theme will intertwine with the original theme to produce further and more beautiful results.

We begin with the observation that the inequality

$$a^2 + b^2 \geq 2ab,$$

on which all the proofs in the preceding sections of this chapter were based [see Sec. 4.2(b)], is a simple consequence of the identity

$$a^2 - 2ab + b^2 = (a - b)^2,$$

which is valid for all real a, b, not merely for nonnegative a, b.

Consider now the product

$$(4.35) \qquad (a^2 + b^2)(c^2 + d^2).$$

We see, upon multiplying out, that this product yields

$$a^2c^2 + b^2d^2 + a^2d^2 + b^2c^2,$$

which is identically what we obtain from expanding the expression

$$(4.36) \qquad (ac + bd)^2 + (bc - ad)^2.$$

Hence we have

$$(4.37) \qquad (a^2 + b^2)(c^2 + d^2) = (ac + bd)^2 + (bc - ad)^2.$$

Since the square $(bc - ad)^2$ is nonnegative, from (4.37) we obtain

$$(4.38) \qquad (a^2 + b^2)(c^2 + d^2) \geq (ac + bd)^2, \qquad \text{for all real } a, b, c, d,$$

a very pretty inequality of great importance throughout much of analysis and mathematical physics. It is called the *Cauchy inequality,* or, more precisely, the two-dimensional version of the Cauchy inequality.†

Furthermore, we see from (4.37) that the sign of equality holds in (4.38) if and only if

$$(4.39) \qquad bc - ad = 0.$$

In this case, we say that the two pairs (a, b) and (c, d) are *proportional* to each other; if $c \neq 0$ and $d \neq 0$, the condition (4.39) can be written as

$$\frac{a}{c} = \frac{b}{d}.$$

(b) *Geometric Interpretation.* On first seeing the identity of the expressions (4.35) and (4.36), the reader should quite naturally and legitimately wonder how in the world anyone would ever stumble upon this result. It strikes him as having been "pulled out of a hat," a piece of mathematical sleight of hand.

It is a tenet of a mathematician's faith that there are no accidents in mathematics. Every result of any significance has an explanation

---

† A generalization of this inequality to expressions occurring in integral calculus was discovered independently by the mathematicians Buniakowski and Schwarz. The name "Cauchy-Schwarz inequality" is sometimes applied to the inequality in the text, but more particularly to its generalization.

which, once grasped, makes the result self-evident. The explanation may not immediately be obvious, and it may not be found for some time. Often the significance of a mathematical theorem becomes clear only when looked at from above—that is to say, from the standpoint of a more advanced theory. But the meaning is always there. This is a vitally important point. Were it not for this, mathematics would degenerate into a collection of unrelated formalisms and parlor tricks.

Frequently, the simplest interpretation of an algebraic result is in terms of a geometric setting. Formulas that seem quite strange and complex become obvious when their geometric origin is laid bare.

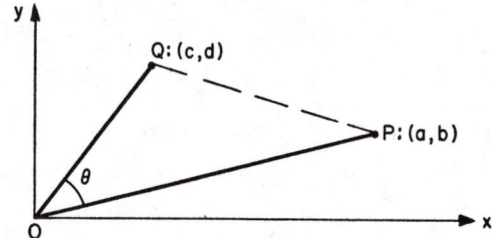

Figure 4.5. Geometric interpretation of the Cauchy inequality

Consider the triangle shown in Fig. 4.5. The lengths of the segments† OP, OQ, and PQ are given by

$$OP = (a^2 + b^2)^{1/2},$$
$$OQ = (c^2 + d^2)^{1/2},$$

and

$$PQ = [(a - c)^2 + (b - d)^2]^{1/2},$$

respectively.

Denote the angle between OP and OQ by θ. By the law of cosines, we have

$$(PQ)^2 = (OP)^2 + (OQ)^2 - 2(OP)(OQ)\cos\theta.$$

Substituting the values of OP, OQ, and PQ, and simplifying, we obtain

(4.40) $$\cos\theta = \frac{ac + bd}{(a^2 + b^2)^{1/2}(c^2 + d^2)^{1/2}}.$$

† The length of a segment XY, often written $\overline{XY}$, will be written XY in this book for reasons of typography.

Since the cosine of an angle must always lie between -1 and $+1$, we must have

(4.41) $0 \leq \cos^2\theta \leq 1$.

Squaring both sides of (4.40) and (4.41), we obtain

$$\cos^2\theta = \frac{(ac + bd)^2}{(a^2 + b^2)(c^2 + d^2)} \leq 1,$$

and finally

$$(a^2 + b^2)(c^2 + d^2) \geq (ac + bd)^2.$$

This is again the two-dimensional version of Cauchy's inequality (4.38), which seemed so magical in its algebraic setting.

Furthermore, we see that there is equality if and only if $\cos^2\theta = 1$, i.e., if and only if θ is a zero or straight angle—that is to say, if and only if the points O, P, and Q lie on the same line. In that case, we must have an equality of slopes; in other words, unless $c = d = 0$,

$$\frac{a}{c} = \frac{b}{d}.$$

(c) *Three-Dimensional Version of the Cauchy Inequality.* An advantage of an interpretation of the foregoing type lies in the fact that we can use geometric intuition to obtain similar results in any number of dimensions.

Turning to three dimensions, let $P:(a_1, a_2, a_3)$ and $Q:(b_1, b_2, b_3)$ be points distinct from the origin $O:(0, 0, 0)$. Then the cosine of the angle θ between OP and OQ is given† by

$$\cos\theta = \frac{a_1 b_1 + a_2 b_2 + a_3 b_3}{(a_1^2 + a_2^2 + a_3^2)^{1/2} (b_1^2 + b_2^2 + b_3^2)^{1/2}},$$

which, together with the fact that $\cos^2\theta \leq 1$, implies the three-dimensional version of the famous inequality of Cauchy:

(4.42) $(a_1^2 + a_2^2 + a_3^2)(b_1^2 + b_2^2 + b_3^2) \geq (a_1 b_1 + a_2 b_2 + a_3 b_3)^2$.

The sign of equality holds if and only if the three points O, P, and

† For a derivation see W. F. Osgood and W. C. Graustein, *Plane and Solid Analytic Geometry,* Macmillan and Co., New York, 1930.

Q are collinear; this is expressed by

$$\frac{a_1}{b_1} = \frac{a_2}{b_2} = \frac{a_3}{b_3},$$

provided the b's are all different from zero.

(d) *The Cauchy-Lagrange Identity and the n-Dimensional Version of the Cauchy Inequality.* A strictly algebraic proof of the three-dimensional Cauchy inequality (4.42) may be obtained by noting that

$$(a_1{}^2 + a_2{}^2 + a_3{}^2)(b_1{}^2 + b_2{}^2 + b_3{}^2) - (a_1 b_1 + a_2 b_2 + a_3 b_3)^2$$

(4.43)
$$= (a_1{}^2 b_2{}^2 + a_2{}^2 b_1{}^2) + (a_1{}^2 b_3{}^2 + a_3{}^2 b_1{}^2) + (a_2{}^2 b_3{}^2 + a_3{}^2 b_2{}^2)$$
$$- 2a_1 b_1 a_2 b_2 - 2a_1 b_1 a_3 b_3 - 2a_2 b_2 a_3 b_3$$

$$= (a_1 b_2 - a_2 b_1)^2 + (a_1 b_3 - a_3 b_1)^2 + (a_2 b_3 - a_3 b_2)^2.$$

The last expression in (4.43) clearly must be nonnegative, since it is a sum of three nonnegative terms. Hence

$$(a_1{}^2 + a_2{}^2 + a_3{}^2)(b_1{}^2 + b_2{}^2 + b_3{}^2) - (a_1 b_1 + a_2 b_2 + a_3 b_3)^2 \geq 0,$$

and thus Cauchy's inequality for three dimensions is proved again. The identity (4.43) yields both the inequality and the case of equality; the last expression in (4.43) is zero if each term vanishes, i.e., if the a's and b's are proportional.

When you study the analytic geometry of three dimensions, you will see that the identity (4.43) is simply the result

$$\cos^2\theta + \sin^2\theta = 1$$

written in another way.

The identity (4.43) can be generalized as follows: For any set of real quantities a_i and b_i ($i = 1, 2, \ldots, n$), we have

(4.44)
$$(a_1{}^2 + a_2{}^2 + \cdots + a_n{}^2)(b_1{}^2 + b_2{}^2 + \cdots + b_n{}^2)$$
$$- (a_1 b_1 + a_2 b_2 + \cdots + a_n b_n)^2$$
$$= (a_1 b_2 - a_2 b_1)^2 + (a_1 b_3 - a_3 b_1)^2 + \cdots + (a_{n-1} b_n - a_n b_{n-1})^2;$$

this is the famous *identity of Cauchy and Lagrange.* From this identity we obtain the n-dimensional version of the Cauchy inequality

(4.45)
$$(a_1{}^2 + a_2{}^2 + \cdots + a_n{}^2)(b_1{}^2 + b_2{}^2 + \cdots + b_n{}^2)$$
$$\geq (a_1 b_1 + a_2 b_2 + \cdots + a_n b_n)^2,$$

valid for all real values of the a_i and b_i.

(e) *An Alternative Proof of the Three-Dimensional Version.* The proofs that we have given above were perfectly satisfactory as far as the results at hand were concerned. However, they do not generalize in such a way as to yield some results that we wish to derive later on. Consequently, we start all over again and introduce a new device.

Let us begin with our basic inequality $(x - y)^2 \geq 0$ in the form

$$(4.46) \qquad \frac{x^2}{2} + \frac{y^2}{2} \geq xy,$$

which is valid for all real x and y, and substitute, in turn,

$$x = \frac{a_1}{(a_1{}^2 + a_2{}^2 + a_3{}^2)^{1/2}}, \qquad y = \frac{b_1}{(b_1{}^2 + b_2{}^2 + b_3{}^2)^{1/2}},$$

then

$$x = \frac{a_2}{(a_1{}^2 + a_2{}^2 + a_3{}^2)^{1/2}}, \qquad y = \frac{b_2}{(b_1{}^2 + b_2{}^2 + b_3{}^2)^{1/2}},$$

then

$$x = \frac{a_3}{(a_1{}^2 + a_2{}^2 + a_3{}^2)^{1/2}}, \qquad y = \frac{b_3}{(b_1{}^2 + b_2{}^2 + b_3{}^2)^{1/2}},$$

where the a_i and b_i are all real quantities. Adding the three inequalities obtained in this way, we find

$$\frac{1}{2}\left(\frac{a_1{}^2 + a_2{}^2 + a_3{}^2}{a_1{}^2 + a_2{}^2 + a_3{}^2}\right) + \frac{1}{2}\left(\frac{b_1{}^2 + b_2{}^2 + b_3{}^2}{b_1{}^2 + b_2{}^2 + b_3{}^2}\right)$$
$$\geq \frac{a_1 b_1 + a_2 b_2 + a_3 b_3}{(a_1{}^2 + a_2{}^2 + a_3{}^2)^{1/2}(b_1{}^2 + b_2{}^2 + b_3{}^2)^{1/2}},$$

which, very obligingly, yields the inequality

$$1 \geq \frac{a_1 b_1 + a_2 b_2 + a_3 b_3}{(a_1{}^2 + a_2{}^2 + a_3{}^2)^{1/2}(b_1{}^2 + b_2{}^2 + b_3{}^2)^{1/2}},$$

equivalent to Cauchy's inequality (4.42) [subsection (c)], as desired. The n-dimensional version (4.45) of Cauchy's inequality may be proved in an analogous manner.

4.5 The Hölder Inequality

We now possess all the tools necessary to fashion one of the most useful inequalities of analysis, the *Hölder inequality*. This states that for any set of nonnegative quantities a_i and b_i $(i = 1, 2, \ldots, n)$, we have

$$(4.47) \qquad (a_1{}^p + a_2{}^p + \cdots + a_n{}^p)^{1/p}(b_1{}^q + b_2{}^q + \cdots + b_n{}^q)^{1/q}$$
$$\geq a_1 b_1 + a_2 b_2 + \cdots + a_n b_n,$$

where p and q are related by the equation

$$(4.48) \qquad \frac{1}{p} + \frac{1}{q} = 1,$$

and $p > 1$.

The case $p = q = 2$ is the Cauchy inequality established in the preceding sections. In the general case, however, we must restrict our attention to nonnegative a_i and b_i, since fractional powers p and q might be involved.

Actually we shall prove (4.47) only for rational numbers p, q; however, the result is valid also for irrational p and q.

We begin with the inequality

$$\frac{a^p}{p} + \frac{b^q}{q} \geq ab,$$

which we established in Sec. 4.3; see (4.33) for rational p, q and non-negative a, b. Then we use the trick employed in Sec. 4.4. We set

$$a = \frac{a_1}{(a_1{}^p + a_2{}^p + \cdots + a_n{}^p)^{1/p}}, \qquad b = \frac{b_1}{(b_1{}^q + b_2{}^q + \cdots + b_n{}^q)^{1/q}},$$

then

$$a = \frac{a_2}{(a_1{}^p + a_2{}^p + \cdots + a_n{}^p)^{1/p}}, \qquad b = \frac{b_2}{(b_1{}^q + b_2{}^q + \cdots + b_n{}^q)^{1/q}},$$

and so on, and add the resulting inequalities. We thus obtain

$$\frac{1}{p}\left(\frac{a_1{}^p + a_2{}^p + \cdots + a_n{}^p}{a_1{}^p + a_2{}^p + \cdots + a_n{}^p}\right) + \frac{1}{q}\left(\frac{b_1{}^q + b_2{}^q + \cdots + b_n{}^q}{b_1{}^q + b_2{}^q + \cdots + b_n{}^q}\right)$$
$$(4.49)$$
$$\geq \frac{a_1 b_1 + a_2 b_2 + \cdots + a_n b_n}{(a_1{}^p + a_2{}^p + \cdots + a_n{}^p)^{1/p}(b_1{}^q + b_2{}^q + \cdots + b_n{}^q)^{1/q}},$$

or, finally, using (4.48), the result stated in the inequality (4.47). Equality can occur in (4.49) if and only if $b_i{}^q/a_i{}^p$ has a common value for all i.

4.6 The Triangle Inequality

We are all familiar with the geometric theorem that the sum of the lengths of two sides of a triangle is greater than or equal to the length of the third side. Let us see what this implies in algebraic terms.

Take the triangle to be situated as shown in Fig. 4.6. Then the geometric inequality

$$OP + PR \geq OR$$

is equivalent to the algebraic *triangle inequality*

(4.50) $$\sqrt{x_1^2 + y_1^2} + \sqrt{x_2^2 + y_2^2} \geq \sqrt{(x_1 + x_2)^2 + (y_1 + y_2)^2}.$$

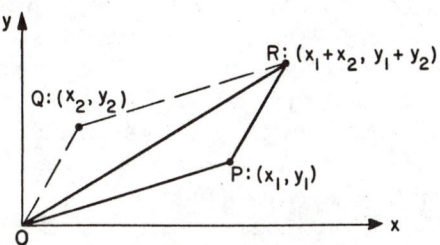

Figure 4.6. The triangle inequality

Can we establish the triangle inequality (4.50) without resorting to geometry? In the one-dimensional case this was done in Sec. 3.8 (see Theorem 3.2), where the inequality was written in the form

$$|x_1| + |x_2| \geq |x_1 + x_2|$$

more often than in the equivalent form

$$\sqrt{x_1^2} + \sqrt{x_2^2} \geq \sqrt{(x_1 + x_2)^2}.$$

Now, the simplest way of proving the two-dimensional triangle inequality (4.50) is to establish the validity of an equivalent inequality. To this end, we square both members of (4.50), obtaining the equivalent inequality

$$x_1^2 + y_1^2 + x_2^2 + y_2^2 + 2\sqrt{x_1^2 + y_1^2}\,\sqrt{x_2^2 + y_2^2}$$
$$\geq (x_1 + x_2)^2 + (y_1 + y_2)^2,$$

which is readily seen to be equivalent to

(4.51) $$\sqrt{x_1^2 + y_1^2}\,\sqrt{x_2^2 + y_2^2} \geq x_1 x_2 + y_1 y_2;$$

but this is a simple consequence of the familiar Cauchy inequality [two-dimensional version, see (4.38)]

$$(4.52) \qquad (x_1{}^2 + y_1{}^2)(x_2{}^2 + y_2{}^2) \geq (x_1x_2 + y_1y_2)^2,$$

and thus the triangle inequality is established.

As in the one-dimensional case (Theorem 3.2), the discussion of the conditions under which the sign of equality holds in the triangle inequality (4.50) is just a bit involved. You will recall that in the Cauchy inequality (4.52) the sign of equality holds if and only if (x_1, y_1) and (x_2, y_2) are proportional, i.e., $x_1 = kx_2$, $y_1 = ky_2$. Now (4.51) is obtained from (4.52) by taking a square root of each member; this is a valid operation since it is the *nonnegative* square root of the left-hand member that was taken. But if $x_1x_2 + y_1y_2 < 0$, that is to say, if $x_1x_2 + y_1y_2$ is the negative square root of the right-hand member $(x_1x_2 + y_1y_2)^2$ of (4.52), then the strict inequality holds in (4.51) even though (x_1, y_1) and (x_2, y_2) are proportional. Thus *the sign of equality holds* in (4.51), and hence also *in the triangle inequality* (4.50), *if and only if* $x_1 = kx_2$ *and* $y_1 = ky_2$ *with a nonnegative constant* k *of proportionality.*†

Geometrically, the foregoing necessary and sufficient condition for the sign of equality to hold in the triangle inequality (4.50) has the interpretation that the points O, P, and Q in Fig. 4.5 be collinear and that P and Q be on the same side of O. Then the triangle OPR collapses. In this case, we say not merely that P and Q are collinear with O, but that *they are on the same ray* issuing from O.

You might check that the foregoing discussion is consistent with the discussion of the corresponding condition in the one-dimensional case $|a| + |b| \geq |a + b|$, namely, that the equality sign holds if and only if a and b are of the same sign.

Our proof of the triangle inequality can be generalized, following the pattern set in the proof of the Hölder inequality, to yield

$$\sqrt{x_1{}^2 + x_2{}^2 + \cdots + x_n{}^2} + \sqrt{y_1{}^2 + y_2{}^2 + \cdots + y_n{}^2}$$
$$\geq \sqrt{(x_1 + y_1)^2 + (x_2 + y_2)^2 + \cdots + (x_n + y_n)^2},$$

valid for all real x_i, y_i, and with the sign of equality holding, as before, if and only if the x_i and the y_i are proportional with a positive factor of proportionality. We shall return to this inequality in Chapter 6,

---

† For example, $(-3, 4)$ and $(6, -8)$ are proportional with a negative constant of proportionality, while $(-3, 4)$ and $(-6, 8)$ have a positive constant of proportionality.

when we consider its geometric significance.

An alternative proof of the triangle inequality, which can be extended to give a more general result, is the following. We write the identity

$$(x_1 + x_2)^2 + (y_1 + y_2)^2$$
$$= x_1(x_1 + x_2) + y_1(y_1 + y_2) + x_2(x_1 + x_2) + y_2(y_1 + y_2)$$

and apply the Cauchy inequality in square-root form [see eq. (4.51)] separately to the two expressions

$$x_1(x_1 + x_2) + y_1(y_1 + y_2)$$

and

$$x_2(x_1 + x_2) + y_2(y_1 + y_2),$$

obtaining

$$(x_1^2 + y_1^2)^{1/2}[(x_1 + x_2)^2 + (y_1 + y_2)^2]^{1/2}$$
$$\geq x_1(x_1 + x_2) + y_1(y_1 + y_2),$$

and

$$(x_2^2 + y_2^2)^{1/2}[(x_1 + x_2)^2 + (y_1 + y_2)^2]^{1/2}$$
$$\geq x_2(x_1 + x_2) + y_2(y_1 + y_2).$$

Adding, we have

$$[(x_1^2 + y_1^2)^{1/2} + (x_2^2 + y_2^2)^{1/2}][(x_1 + x_2)^2 + (y_1 + y_2)^2]^{1/2}$$
$$\geq (x_1 + x_2)^2 + (y_1 + y_2)^2.$$

Dividing through by the common factor $[(x_1 + x_2)^2 + (y_1 + y_2)^2]^{1/2}$, we get

$$(x_1^2 + y_1^2)^{1/2} + (x_2^2 + y_2^2)^{1/2} \geq [(x_1 + x_2)^2 + (y_1 + y_2)^2]^{1/2}.$$

Thus we have again established the triangle inequality (4.50).

Returning to the point where the Cauchy inequality was applied in square-root form, we see again that equality occurs if and only if

$$x_1 = kx_2, \qquad y_1 = ky_2$$

for some nonnegative factor of proportionality k—that is to say, if and only if the three points O, P, and Q are collinear, with P and Q on the same side of O.

4.7 The Minkowski Inequality

We now possess all the tools and devices to establish another famous inequality, that due to Minkowski. It asserts that for any set of nonnegative† quantities x_1, y_1, x_2, y_2, and any $p > 1$, we have

$$(4.53) \quad (x_1{}^p + y_1{}^p)^{1/p} + (x_2{}^p + y_2{}^p)^{1/p}$$
$$\geq [(x_1 + x_2)^p + (y_1 + y_2)^p]^{1/p}.$$

The triangle inequality is the special case $p = 2$.

The proof of Minkowski's inequality is similar to that just given for the triangle inequality, with the difference that the more general Hölder inequality (Sec. 4.5) is used in place of Cauchy's inequality. Write the identity

$$(x_1 + x_2)^p + (y_1 + y_2)^p = [x_1(x_1 + x_2)^{p-1} + y_1(y_1 + y_2)^{p-1}]$$
$$+ [x_2(x_1 + x_2)^{p-1} + y_2(y_1 + y_2)^{p-1}]$$

and apply Hölder's inequality to each of the terms separately. The results are

$$(x_1{}^p + y_1{}^p)^{1/p}[(x_1 + x_2)^{(p-1)q} + (y_1 + y_2)^{(p-1)q}]^{1/q}$$
$$\geq x_1(x_1 + x_2)^{p-1} + y_1(y_1 + y_2)^{p-1}$$

and

$$(x_2{}^p + y_2{}^p)^{1/p}[(x_1 + x_2)^{(p-1)q} + (y_1 + y_2)^{(p-1)q}]^{1/q}$$
$$\geq x_2(x_1 + x_2)^{p-1} + y_2(y_1 + y_2)^{p-1}.$$

Since $1/p + 1/q = 1$, we see that $(p - 1)q = p$. Adding, we obtain

$$[(x_1 + x_2)^p + (y_1 + y_2)^p]^{1/q}[(x_1{}^p + y_1{}^p)^{1/p} + (x_2{}^p + y_2{}^p)^{1/p}]$$
$$\geq (x_1 + x_2)^p + (y_1 + y_2)^p;$$

and then dividing by $[(x_1 + x_2)^p + (y_1 + y_2)^p]^{1/q}$, we get

$$(x_1{}^p + y_1{}^p)^{1/p} + (x_2{}^p + y_2{}^p)^{1/p} \geq [(x_1 + x_2)^p + (y_1 + y_2)^p]^{1-1/q}$$

which, since $1 - 1/q = 1/p$, is the stated Minkowski inequality (4.53).

The sign of equality holds in the Minkowski inequality if and only if it holds in the Hölder inequality [by means of which (4.53) was proved], that is, if and only if the points (x_1, y_1) and (x_2, y_2) [which are in the first quadrant] are collinear.

---

† The restriction to nonnegative quantities is necessary because, in general, fractional powers p and q are involved.

As you might surmise from the foregoing generalizations of the Cauchy, Hölder, and triangle inequalities, the general Minkowski inequality for sets of n nonnegative numbers

$$x_1, x_2, \ldots, x_n \quad \text{and} \quad y_1, y_2, \ldots, y_n$$

is

$$[x_1^p + x_2^p + \cdots + x_n^p]^{1/p} + [y_1^p + y_2^p + \cdots + y_n^p]^{1/p}$$
$$\geq [(x_1 + y_1)^p + (x_2 + y_2)^p + \cdots + (x_n + y_n)^p]^{1/p}$$

for $p \geq 1$. The inequality sign is reversed for $p < 1$.

4.8 Absolute Values and the Classical Inequalities

The arithmetic-mean–geometric-mean inequality, the Cauchy inequality, the Hölder inequality, the triangle inequality, and the Minkowski inequality are the classical inequalities of mathematical analysis. For ready reference, they are collected in Table 3.

TABLE 3. The Classical Inequalities for Nonnegative Values

Name	Inequality
Arithmetic-mean–geometric-mean inequality	$\dfrac{a_1 + a_2 + \cdots + a_n}{n} \geq (a_1 a_2 \cdots a_n)^{1/n}$
Cauchy inequality	$(a_1^2 + a_2^2 + \cdots + a_n^2)^{1/2}(b_1^2 + b_2^2 + \cdots + b_n^2)^{1/2}$ $\geq a_1 b_1 + a_2 b_2 + \cdots + a_n b_n$
Hölder inequality	$(a_1^p + a_2^p + \cdots + a_n^p)^{1/p}(b_1^q + b_2^q + \cdots + b_n^q)^{1/q}$ $\geq a_1 b_1 + a_2 b_2 + \cdots + a_n b_n$
Triangle inequality	$(a_1^2 + a_2^2 + \cdots + a_n^2)^{1/2} + (b_1^2 + b_2^2 + \cdots + b_n^2)^{1/2}$ $\geq [(a_1 + b_1)^2 + (a_2 + b_2)^2 + \cdots + (a_n + b_n)^2]^{1/2}$
Minkowski inequality	$(a_1^p + a_2^p + \cdots + a_n^p)^{1/p} + (b_1^p + b_2^p + \cdots + b_n^p)^{1/p}$ $\geq [(a_1 + b_1)^p + (a_2 + b_2)^p + \cdots + (a_n + b_n)^p]^{1/p}$

The inequalities are valid for any nonnegative values $a_1, a_2, \ldots, a_n$ and $b_1, b_2, \ldots, b_n$; for arbitrary $p > 1$; and for $1/p + 1/q = 1$. The Cauchy inequality and the triangle inequality constitute the special case $p = 2$ of the Hölder inequality and the Minkowski inequality, respectively. The sign of equality holds if and only if all the a_i are equal, for the arithmetic-mean–geometric-mean inequality; the sets (a_i^p), (b_i^q) are proportional, for the Hölder inequality; and the sets (a_i), (b_i) are proportional, for the other inequalities.

The foregoing inequalities are primarily concerned with nonnegative values. However, the absolute value of any real number is nonnegative, so that the inequalities apply, in particular, to the absolute values of arbitrary real numbers. This observation can be extended somewhat by means of the result in Theorem 3.2, which states that a sum of absolute values is greater than or equal to the absolute value of the corresponding sum; thus, relative to the Minkowski inequality, we have

$$|a_i| + |b_i| \geq |a_i + b_i|,$$

the sign of equality holding if and only if a_i and b_i are of the same sign. The extended inequalities are exhibited in Table 4.

TABLE 4. The Classical Inequalities for Arbitrary Values

Name	Inequality
Arithmetic-mean–geometric-mean inequality	$\dfrac{|a_1| + |a_2| + \cdots + |a_n|}{n} \geq |a_1 a_2 \cdots a_n|^{1/n}$
Cauchy inequality	$(a_1^2 + a_2^2 + \cdots + a_n^2)^{1/2}(b_1^2 + b_2^2 + \cdots + b_n^2)^{1/2}$ $\geq a_1 b_1 + a_2 b_2 + \cdots + a_n b_n$
Hölder inequality	$(|a_1|^p + |a_2|^p + \cdots + |a_n|^p)^{1/p}$ $\cdot (|b_1|^q + |b_2|^q + \cdots + |b_n|^q)^{1/q}$ $\geq a_1 b_1 + a_2 b_2 + \cdots + a_n b_n$
Triangle inequality	$(a_1^2 + a_2^2 + \cdots + a_n^2)^{1/2} + (b_1^2 + b_2^2 + \cdots + b_n^2)^{1/2}$ $\geq [(a_1 + b_1)^2 + (a_2 + b_2)^2 + \cdots + (a_n + b_n)^2]^{1/2}$
Minkowski inequality	$(|a_1|^p + |a_2|^p + \cdots + |a_n|^p)^{1/p}$ $+ (|b_1|^p + |b_2|^p + \cdots + |b_n|^p)^{1/p}$ $\geq [|a_1 + b_1|^p + |a_2 + b_2|^p + \cdots + |a_n + b_n|^p]^{1/p}$

For these inequalities involving arbitrary real values $a_1, a_2, \ldots, a_n$ and $b_1, b_2, \ldots, b_n$, the sign of equality holds if and only if all the $|a_i|$ are equal, for the arithmetic-mean–geometric-mean inequality; the sets $(|a_i|^p)$, $(|b_i|^q)$ are proportional and each $a_i b_i$ nonnegative, for the Hölder inequality; and the sets (a_i), (b_i) are nonnegatively proportional, for the other inequalities.

As an application, for given sets of values $(x_1, x_2, \ldots, x_n)$, $(y_1, y_2, \ldots, y_n)$, and $(z_1, z_2, \ldots, z_n)$, let

$$a_i = z_i - y_i \quad \text{and} \quad b_i = y_i - x_i.$$

Then

$$a_i + b_i = z_i - y_i + y_i - x_i = z_i - x_i,$$

and the extended Minkowski inequality yields

$$(|z_1 - y_1|^p + |z_2 - y_2|^p + \cdots + |z_n - y_n|^p)^{1/p}$$
$$+ (|y_1 - x_1|^p + |y_2 - x_2|^p + \cdots + |y_n - x_n|^p)^{1/p}$$
$$\geq (|z_1 - x_1|^p + |z_2 - x_2|^p + \cdots + |z_n - x_n|^p)^{1/p};$$

equality holds if and only if the sets $(z_1 - y_1, z_2 - y_2, \ldots, z_n - y_n)$ and $(y_1 - x_1, y_2 - x_2, \ldots, y_n - x_n)$ are proportional with a non-negative constant of proportionality.

4.9 Symmetric Means

You might think, by this time, that the subject of inequalities has been pretty well exhausted. Actually, just the opposite is true. All that has gone before represents the barest scratching of the surface. The inequalities that we have established can be generalized in countless ways, and then these generalizations in turn can be generalized. They can be combined in many ingenious ways to give many further results, as we shall indicate in this section and in Sec. 4.10.

For example, one extension of the arithmetic-mean–geometric-mean inequality that is rather interesting is the following: Form the three means,

$$m_1 = \frac{x_1 + x_2 + x_3}{3},$$

$$m_2 = \left(\frac{x_1 x_2 + x_1 x_3 + x_2 x_3}{3}\right)^{1/2},$$

$$m_3 = (x_1 x_2 x_3)^{1/3}.$$

Then the continued inequality

$$m_1 \geq m_2 \geq m_3$$

is valid for all nonnegative quantities x_1, x_2, and x_3. We leave the proof of this for the reader. This, in turn, is a special case of the analogous result valid for any n nonnegative quantities, in which there are n mean values starting with the arithmetic mean and ending with the geometric.

4.10 The Arithmetic-Geometric Mean of Gauss

Let a and b be two nonnegative quantities, and consider the quantities a_1 and b_1 defined in terms of a and b by

$$a_1 = \frac{a+b}{2}, \qquad b_1 = \sqrt{ab}.$$

By our basic Axiom II, page 7, a_1 and b_1 must also be nonnegative. If we assume that $a > b$, we have

$$a_1 < \frac{a+a}{2} = a, \qquad b_1 > \sqrt{b^2} = b.$$

Furthermore, by the arithmetic-mean–geometric-mean inequality,

$$a_1 > b_1.$$

Let us now repeat this process, introducing a_2 and b_2 by means of the formulas

$$a_2 = \frac{a_1 + b_1}{2}, \qquad b_2 = \sqrt{a_1 b_1}.$$

Then, by the same reasoning as above,

$$a > a_1 > a_2,$$
$$b < b_1 < b_2,$$

and

$$a_2 > b_2.$$

Let us now continue this process, defining

$$a_3 = \frac{a_2 + b_2}{2}, \qquad b_3 = \sqrt{a_2 b_2},$$

then a_4, b_4, and generally a_n, b_n by the recurrence relationships

$$(4.54) \qquad a_n = \frac{a_{n-1} + b_{n-1}}{2}, \qquad b_n = \sqrt{a_{n-1}b_{n-1}}.$$

After the kth step of our process, we have numbers $a_1, a_2, \ldots, a_k$ and $b_1, b_2, \ldots, b_k$ satisfying the inequalities

$$a > a_1 > a_2 > \cdots > a_k > b_k > b_{k-1} > \cdots > b_1 > b.$$

For example, if $a = 4$ and $b = 1$, then the first few a's and b's are represented on the line shown in Fig. 4.7. We observe that b is the smallest and a the largest of all these numbers, and that all the b's are smaller than all the a's.

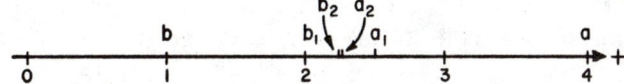

Figure 4.7. The arithmetic-geometric mean of Gauss

We need not stop after k steps. Suppose we continue defining more and more a's and b's by the relationships (4.54). Each pair a_i, b_i, so defined, is sandwiched between the previous pair a_{i-1}, b_{i-1}. Thus it is plausible that the quantities a_n, which decrease as n gets larger but which must always remain bigger than each b_i, approach some fixed value A; similarly, the b_n, which increase as n gets larger but which remain smaller than each a_i, approach a fixed value B. The reader familiar with the concept of limit will readily see that A is the limit of the infinite sequence of numbers $\{a_n\}$ and B is the limit of the infinite sequence $\{b_n\}$.

Moreover, the difference $a_n - b_n$ becomes rapidly smaller as n increases. We can show that, at each step, this difference is less than half of what it was at the previous step. Just write, using (4.54),

$$(4.55) \qquad a_{n+1} - b_{n+1} = \frac{a_n + b_n}{2} - \sqrt{a_n b_n} = \frac{a_n - 2\sqrt{a_n b_n} + b_n}{2}.$$

Since $\sqrt{b_n} < \sqrt{a_n}$, we have

$$2b_n < 2\sqrt{a_n b_n};$$

adding $a_n - b_n - 2\sqrt{a_n b_n}$ to both sides of this inequality, we obtain

$$a_n - 2\sqrt{a_n b_n} + b_n < a_n - b_n$$

which, when applied to (4.55), yields

$$a_{n+1} - b_{n+1} < \frac{1}{2}(a_n - b_n)$$

as we set out to show.

Since the a's and b's approach each other in this manner, their limits A and B must coincide. So

$$A = B,$$

and this common limit depends only on the numbers a and b with which we started; in mathematical language, A is a function of a and b. It was shown by Gauss that this function is not just a curiosity but is rather basic in analysis; it may be used to found the branch of mathematics called the "theory of elliptic functions."

CHAPTER FIVE

Maximization and Minimization Problems

5.1 Introduction

Now we shall demonstrate how the inequalities that were derived in the preceding chapter can be used to treat an important and fascinating set of problems. These are *maximization* and *minimization* problems.

In the study of algebra and trigonometry, the problems you encountered were all of the following general nature: Given certain initial data, and certain operations that were to be performed, you were to determine the outcome. Or, conversely, given the operations that were performed and the outcome, you were to determine what the initial data must have been.

For example, you met three workmen, eager and industrious *A*, plodding and conscientious *B*, and downright lazy *C*. They were put to work digging ditches, constructing swimming pools, or building houses, and the problem was always that of determining how long it would take the three of them together to perform the job, given their individual efficiency ratings—or, given the time required for all three together, and the rates of effort of two of the three workers, it was required to find the rate of the third.

Sometimes you were given triangles in which two sides and an included angle were known, and sometimes you were given three sides of the triangle. In each case, the problem was that of determining the remaining sides and angles.

The foregoing kinds of problems are simple versions of what may be called *descriptive problems*.

Here, we wish to consider quite different types of problems. We shall consider situations in which there are many ways of proceeding, and for which the problem is that of determining *optimal choices*. Questions of this nature occur in all branches of science and constitute one of the most important applications of mathematical analysis. Furthermore, much of physics and engineering is dominated by principles asserting that physical processes occur in nature as if some quantity, such as time or energy, were being minimized or maximized.

Many processes of this sort can be treated in routine fashion by that magic technique of Fermat, Leibniz, and Newton, the calculus. Supposedly, in the 17th century, men who knew calculus were pointed out in the street as possessing this extraordinary knowledge. Today it is a subject that can be taught in high school and college, and requires no exceptional aptitude.

When you take a course in calculus, you will find that many of the problems treated here by algebraic processes can also be solved quite readily by means of calculus. It will then be amusing to solve them mentally by means of the techniques presented here, but to follow the formalism of calculus in order to check your answer.

Each technique, calculus as well as the theory of inequalities, has its own advantages and disadvantages as far as applications to maximization and minimization problems are concerned. It is rather characteristic of mathematics that there should be a great overlapping of techniques in the solution of any particular problem. Usually, a problem that can be solved in one way can also be solved in a number of other ways.

5.2 The Problem of Dido

According to legend, the city of Carthage was founded by Dido, a princess from the land of Tyre. Seeking land for this new settlement, she obtained a grudging concession from the local natives to occupy as much land as could be encompassed by a cowhide. Realizing that, taken literally, this would result in a certain amount of overcrowding, she very ingeniously cut the skin into thin strips and then strung them out so as to surround a much larger territory than the one-cowhide area that the natives had intended.

The mathematical problem that she faced was that of determining the closed curve of fixed perimeter that would surround the greatest area; see Fig. 5.1.

In its general form, the problem is much too complicated for us to handle. As a matter of fact, at our present level we don't even know how to formulate it in precise terms, since we have no way of expressing what we mean either by the perimeter of an arbitrary curve or by the area enclosed by this curve. Calculus suggests definitions and provides analytic expressions for these quantities, and a still more advanced part of mathematics called the *calculus of variations* provides a solution to the problem.

As one might guess, the optimal curve in Dido's problem is a circle. Let us here, however, concentrate on certain simple versions of the problem that we can easily handle by means of the fundamental results we have already obtained in the theory of inequalities. If you would like to pursue the study still further, you should read the interesting tract *Geometric Inequalities* by N. D. Kazarinoff, also in this series.

5.3 A Simplified Version of Dido's Problem

Let us suppose, for various practical considerations, that Dido was constrained to select a rectangular plot of land, as indicated in Fig. 5.2.

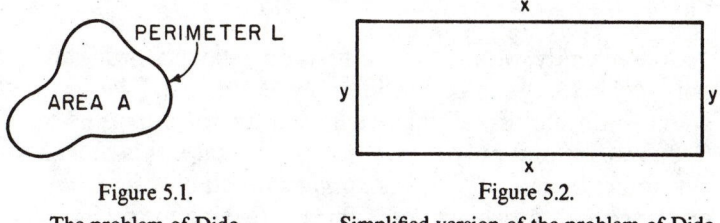

Figure 5.1.	Figure 5.2.
The problem of Dido	Simplified version of the problem of Dido

With the lengths of the sides of the rectangle designated by x and y, respectively, we see that the length of its perimeter is given by the algebraic expression

$$(5.1) \qquad L = 2x + 2y,$$

and its area is represented by the formula

$$(5.2) \qquad A = xy.$$

Since x and y represent lengths, they necessarily are nonnegative quantities. The fact that the perimeter $2x + 2y$ is equal to L implies that x and y must satisfy the inequalities

$$(5.3) \qquad \frac{L}{2} \geq x \geq 0, \qquad \frac{L}{2} \geq y \geq 0.$$

It is clear, then, that the area $A = xy$ cannot be arbitrarily large. Indeed, from the inequalities (5.3) and the expression (5.2) for the area, we see by Theorem 2.5 of Chapter 2 that A cannot exceed the value $L^2/4$; that is, we have

$$(5.4) \qquad \frac{L^2}{4} \geq xy = A.$$

How are we going to determine the dimensions that yield the maximum area?

Referring to the inequality connecting the arithmetic mean and geometric mean of two quantities, we observe that

$$(5.5) \qquad \frac{x+y}{2} \geq \sqrt{xy}$$

for all nonnegative numbers x and y. Since, in this case, we have $x + y = L/2$, the inequality (5.5) yields the relationship

$$(5.6) \qquad \frac{L}{4} \geq \sqrt{xy} \quad \text{or} \quad \frac{L^2}{16} \geq xy = A.$$

Consequently, we see that our first rough bound for the area, $L^2/4$, obtained in (5.4), can be considerably "tightened." We can, however, go much further. Recall that we have established that the equality in (5.6) holds if and only if $x = y$. In our case, this means that the new upper bound, $L^2/16$, is *attained* if and only if $x = y$. For all other nonnegative choices of x and y satisfying the relationship (5.1), the area is less than $L^2/16$.

What does the foregoing analysis tell us? It tells us that in no case can the area of a rectangle with perimeter L exceed the quantity $L^2/16$, and that this maximum value is actually reached if and only if we choose the sides equal to each other, and therefore equal to the quantity $L/4$.

Thus, by purely algebraic means we have achieved a proof of the well-known and rather intuitively obvious fact that *the rectangle of greatest area for a given perimeter is a square.*

5.4 The Reverse Problem

Let us now consider the problem of determining the rectangle of minimum perimeter that encloses a fixed area. This is a *dual* or *reverse* problem to the original.

Returning to the equations (5.1) and (5.2), we see that we now wish to determine the nonnegative values of x and y that make the expression $2x + 2y$ a minimum, while preserving a fixed value A for the product xy.

As might be expected, the same inequality between the arithmetic and geometric means yields the solution to this problem. From the relationship (5.5) above, we know that

$$\frac{(x + y)^2}{4} \geq xy = A.$$

Hence we see that

$$x + y \geq 2\sqrt{A},$$

and that

$$L = 2x + 2y \geq 4\sqrt{A}.$$

Consequently, we can assert that the perimeter must be at least as great as $4\sqrt{A}$ and, furthermore, that this minimum value is actually attained if and only if $x = y = \sqrt{A}$. Hence, once again, the optimal rectangular shape must be that of a square.

This reciprocal relationship between the solutions of the two problems is no accident. Usually, in the study of variational problems of this type, the solution of one problem automatically yields the solution of the dual problem as well. For a proof of this duality principle, see, e.g., N. D. Kazarinoff's *Geometric Inequalities* mentioned earlier.

5.5 The Path of a Ray of Light

Suppose we wish to determine the path of a ray of light going from a point P to a point Q by way of reflection in a plane surface, as shown in Fig. 5.3. Actually, the problem as here stated is a three-dimensional one, but an extension of the following analysis shows that the ray must travel in the plane through P and Q and perpendicular to the reflecting plane.

We assume that the medium is homogeneous, so that a ray of light travels at constant speed. How shall we determine the point R and the paths *PR* and *RQ?* Let us invoke *Fermat's principle,* which states that the total length of time required must be a minimum for all possible choices of the point *R*. Again, because the medium is homogeneous, this means that the paths *PR* and *RQ* are straight lines and that *R* is so located that the length *PR + RQ* is a minimum.

Let the coordinates of *P, Q* be $(0, a)$, (q, b) respectively, and let *r* be the unknown distance *OR*. Then

$$OP = a, \qquad TQ = b, \qquad OR = r, \qquad RT = q - r,$$

and

$$PR = \sqrt{a^2 + r^2}, \qquad RQ = \sqrt{b^2 + (q - r)^2};$$

see Fig. 5.4. Our problem is to determine *r* so that *PR + RQ* is a minimum, i.e., so that

$$\sqrt{a^2 + r^2} + \sqrt{b^2 + (q - r)^2}$$

is a minimum.

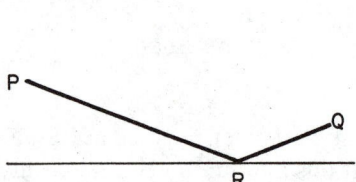

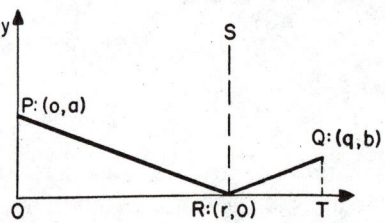

Figure 5.3. Path of a reflected ray of light   Figure 5.4. Determination of the point *R*

We apply the triangle inequality as follows:

$$\sqrt{a^2 + r^2} + \sqrt{b^2 + (q - r)^2} \geq \sqrt{(a + b)^2 + (r + q - r)^2}$$
$$= \sqrt{(a + b)^2 + q^2}.$$

Thus the distance traveled cannot be less than the fixed amount $\sqrt{(a + b)^2 + q^2}$; and this minimum value is achieved precisely if the sets (a, r) and $(b, q - r)$ are proportional with a positive constant of proportionality, that is, if

$$(5.7) \qquad \frac{a}{r} = \frac{b}{q - r} > 0.$$

Observe what the condition (5.7) means geometrically. It means that the right triangles *ORP* and *TRQ* are similar and that, since *b* is positive, $q - r$ is positive so that *R* falls between *O* and *T*. From the similarity of the right triangles, we conclude that the angles *ORP* and *TRQ* are equal. Therefore, their complements $\angle SRP$ and $\angle SRQ$

are also equal. Thus Fermat's principle has enabled us to deduce the famous result that *the angle of incidence is equal to the angle of reflection,*

$$\angle SRP = \angle SRQ,$$

and the more obvious fact that R lies between O and T.

The above principle of reflection can also be proved by purely geometric considerations. If R' in Fig. 5.5 denotes any point on the x-axis different from R, and if Q' is the point $(q, -b)$, then

$$PR + RQ = PR + RQ' = PQ' < PR' + R'Q' = PR' + R'Q,$$

so that the point R is again seen to minimize the distance $PR + RQ$.

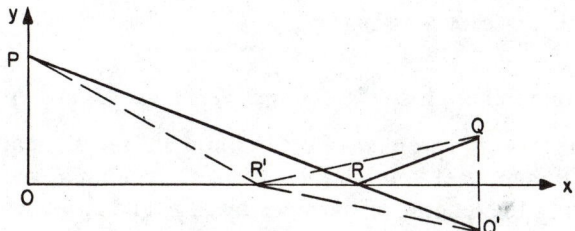

Figure 5.5. Geometric determination of the point R

Suppose a plane divides two homogeneous media M_1 and M_2 of different density so that in M_1 light rays travel with velocity v_1 along straight lines and in M_2 they travel with velocity v_2 along straight lines. We now wish to determine the least time-consuming path from a point P in M_1 to a point Q in M_2; see Fig. 5.6. Again we have

$$PR = \sqrt{a^2 + r^2}, \qquad RQ = \sqrt{b^2 + (q - r)^2}$$

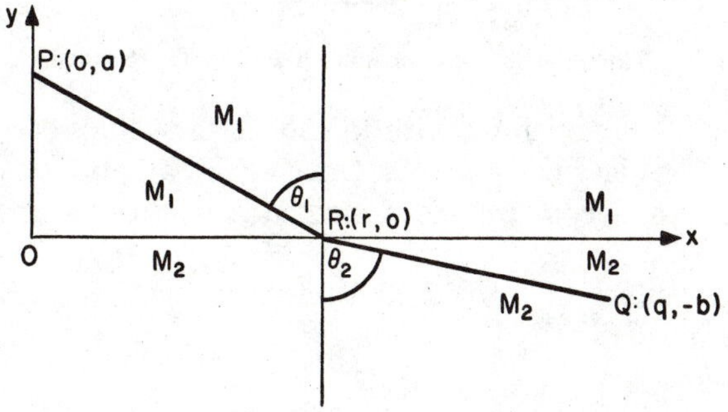

Figure 5.6. A refracted ray of light

and wish to minimize the time (distance/velocity)

$$t = \frac{\sqrt{a^2 + r^2}}{v_1} + \frac{\sqrt{b^2 + (q - r)^2}}{v_2}.$$

While this is an easy problem in differential calculus, the minimum value does not seem to be obtainable directly from our elementary inequalities. The minimum value of t is obtained if the path from P to Q is a broken straight line PRQ (see Fig. 5.6) located so that the angles θ_1 and θ_2 (between PR and the normal to the plane and RQ and the normal to the plane, respectively) satisfy the relation

$$\frac{\sin\theta_1}{\sin\theta_2} = \frac{v_1}{v_2},$$

This relation is known as *Snell's law of refraction.*

5.6 Simplified Three-dimensional Version of Dido's Problem

Consider now the problem of determining the rectangular box that encloses the greatest volume for a fixed surface area; see Fig. 5.7. Designating the lengths of the sides by x, y, and z, we see that the volume is given by the expression

$$V = xyz,$$

while the surface area is represented by the formula

$$A = 2xy + 2xz + 2yz.$$

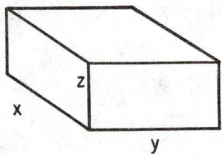

Figure 5.7. Three-dimensional version of the simplified problem of Dido

Given the value of A, we wish to choose the values of x, y, and z that make V as large as possible. Once again, an application of the arithmetic-mean–geometric-mean inequality will resolve the problem for us.

Considering xy, xz, and yz as three nonnegative quantities, we obtain the inequality

(5.8) $$\frac{xy + xz + yz}{3} \geq [(xy)(xz)(yz)]^{1/3} = (xyz)^{2/3}.$$

We know that equality can occur if and only if

$$xy = xz = yz,$$

a relationship that holds if and only if

$$x = y = z.$$

Since

$$xy + xz + yz = \frac{A}{2},$$

the inequality (5.8) yields

$$\frac{A}{6} \geq (xyz)^{2/3} = V^{2/3} \quad \text{or} \quad \left(\frac{A}{6}\right)^{3/2} \geq V.$$

It follows that the volume of the rectangular box with surface area A is less than or equal to $(A/6)^{3/2}$, and that this value is attained if and only if

$$x = y = z = \left(\frac{A}{6}\right)^{1/2}.$$

Hence, the rectangular box of maximum volume for a given surface area is a cube; and, dually, the rectangular box of minimum surface area that encloses a given volume is a cube. Once again, we observe a reciprocal relationship between two problems.

Exercises

1. Show that, if the sum of the lengths of the twelve edges of a rectangular box is a fixed value E, then the surface area A of the box is at most $E^2/24$, and that the box is a cube if and only if $E^2/24 = A$.

2. State and prove the reverse of the result expressed in Exercise 1.

3. Of all rectangles having the same length of diagonal, determine which has the greatest perimeter and determine which has the greatest area. (Use the results in Exercise 3 on page 24 or Exercise 3 on page 62.)

5.7 Triangles of Maximum Area for a Fixed Perimeter

Let us now consider the problem of determining the *triangle* of maximum area for a given perimeter. In Fig. 5.8, let s denote half

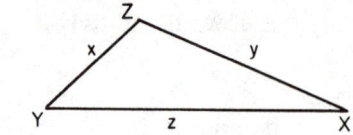

Figure 5.8. The problem of Dido for a triangle

the length of the perimeter of the triangle shown; that is, let

$$s = \frac{x + y + z}{2}.$$

As is well known, the area A of the triangle can be expressed by the formula

$$A = [s(s - x)(s - y)(s - z)]^{1/2}.$$

We wish to find the maximum value of the area as x, y, and z vary over all positive values for which the relationship

$$2s = x + y + z$$

holds, where the value s is fixed.

Once again, the arithmetic-mean–geometric-mean inequality yields the solution in a very simple way. For the three nonnegative values $s - x$, $s - y$, and $s - z$, we have

$$[(s - x)(s - y)(s - z)]^{1/3} \leq \frac{(s - x) + (s - y) + (s - z)}{3}$$

$$\leq \frac{3s - (x + y + z)}{3} = \frac{3s - 2s}{3} = \frac{s}{3}.$$

Hence,

(5.9) $$(s - x)(s - y)(s - z) \leq \left(\frac{s}{3}\right)^3.$$

From (5.9), by easy steps, we obtain the result

$$A = [s(s - x)(s - y)(s - z)]^{1/2} \leq \left[s\left(\frac{s}{3}\right)^3\right]^{1/2} = \left(\frac{s^4}{3^3}\right)^{1/2} = \frac{s^2}{3\sqrt{3}}.$$

There is equality if and only if

$$s - x = s - y = s - z,$$

that is, if and only if $x = y = z$. Consequently, we can assert:

THEOREM 5.1. *Of all triangles with a fixed perimeter, the equilateral triangle maximizes the area.*

Observe that all the results we have so far obtained in this chapter seem to indicate that symmetry and optimal behavior are closely interrelated. Perhaps the profound and aesthetic observation of Keats, "Beauty is truth, truth beauty," sums this up best.

Exercises

1. Let s denote half the length of the perimeter and A the area of the triangle having sides of length x, y, and z, so that

$$s = \frac{x + y + z}{2}$$

and

$$A^2 = s(s - x)(s - y)(s - z).$$

For the special case of an isosceles triangle, with $y = z$, let the area be denoted by I; and for the special case of an equilateral triangle, with $x = y = z$, let the area be denoted by E. Show that

$$I^2 = \frac{s}{4} x^2(s - x)$$

and

$$E^2 = \frac{s^4}{27}.$$

2. With the notation of Exercise 1, show that, for triangles of equal perimeter s,

$$E^2 - I^2 = \frac{s}{27}\left(s + 3x\right)\left(s - \frac{3x}{2}\right)^2$$

and

$$I^2 - A^2 = \frac{s}{4}(s - x)(y - z)^2,$$

where x is the length of the base of the isosceles triangle and also the length of one side of the general triangle.

3. Using the formulas in Exercise 2, show that

$$E^2 - I^2 \geq 0$$

and

$$I^2 - A^2 \geq 0.$$

Under what circumstances do the signs of equality hold?

4. Using one of the inequalities in Exercise 3, show that, if the perimeter and one side of a triangle are given, then the area is maximized by making the triangle isosceles.

5. Using one of the inequalities in Exercise 3, show that of all isosceles triangles with a given perimeter, the equilateral triangle maximizes the area.

6. Using the inequalities in Exercise 3, explain how the formula

$$A = \sqrt{E^2 - (E^2 - I^2) - (I^2 - A^2)}$$

exhibits A in terms of nonnegative expressions, and how the results stated in Exercise 4, Exercise 5, and Theorem 5.1 follow from this formula.

7. Of all right triangles having the same length of hypotenuse, determine which has the greatest altitude to the hypotenuse.

5.8 The Wealthy Football Player

Consider now a problem of a similar but slightly more complex type.

A football player, having parlayed his athletic prowess on the gridiron into a position in Wall Street and thence by natural stages into the possession of a variety of stocks and bonds of his own, became, in the course of time, quite wealthy. Being of a sentimental turn of mind, as retired football players are wont, he stipulated in his last will and testament that he be buried in a vault shaped like a giant football. His executors, honoring his last request, were faced with the problem of carrying out this assignment in the most economical fashion.

After some thought, they decided that the mathematical problem most closely approximating the physical situation that they faced was that of *enclosing a rectangular box of given dimensions within an ellipsoid of the least possible volume.*

Following the leads contained in the foregoing problems, they began with the *reverse* problem, i.e., that of *inscribing a rectangular box of greatest volume in a given ellipsoid.*

According to solid analytic geometry, an ellipsoid with center at the origin and axes along the coordinate axes has an equation of the form

(5.10) $$\frac{x^2}{a^2} + \frac{y^2}{b^2} + \frac{z^2}{c^2} = 1,$$

where $2a$, $2b$, $2c$ represent the lengths of the axes of the ellipsoid. Now, it is intuitively clear that the inscribed rectangular box will have its center at the origin and its sides parallel to the axes. Thus, if one vertex of the box is at the point (x, y, z) on the surface of the ellipsoid (see Fig. 5.9 which shows one eighth of the ellipsoid), then the other seven vertices must be located symmetrically. This means that the

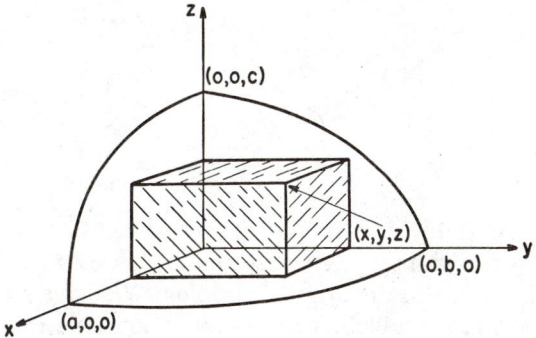

Figure 5.9. One octant of a box in an ellipsoid

coordinates of the other seven vertices of the box are given by

$$(-x, y, z), \quad (x, -y, z), \quad (x, y, -z), \quad (-x, -y, z),$$
$$(x, -y, -z), \quad (-x, y, -z), \quad (-x, -y, -z).$$

Since the sides of the box have lengths $2x$, $2y$, $2z$, it follows that the volume of the box is given by the expression

$$(5.11) \qquad\qquad V = 8xyz.$$

The problem that confronts us is that of maximizing the expression (5.11) subject to the condition that the quantities x, y, and z satisfy eq. (5.10).

Once again, this problem is easily resolved by use of the arithmetic-mean–geometric-mean inequality. We have

$$(5.12) \qquad \frac{1}{3}\!\left(\frac{x^2}{a^2} + \frac{y^2}{b^2} + \frac{z^2}{c^2}\right) \geq \left(\frac{x^2}{a^2} \cdot \frac{y^2}{b^2} \cdot \frac{z^2}{c^2}\right)^{1/3}$$

Using the fact that x, y, and z satisfy eq. (5.10), and using the formula (5.11) for V, we see that the inequality (5.12) yields

$$\frac{1}{3} \geq \frac{V^{2/3}}{4(a^2b^2c^2)^{1/3}}$$

or

$$\frac{4}{3}(abc)^{2/3} \geq V^{2/3}.$$

It follows that the volume of the box is at most $8abc/3\sqrt{3}$; furthermore, by the condition for equality in (5.12), this value is attained if and only if

$$\frac{x^2}{a^2} = \frac{y^2}{b^2} = \frac{z^2}{c^2} = \frac{1}{3}$$

or

(5.13) $$x = \frac{a}{\sqrt{3}}, \qquad y = \frac{b}{\sqrt{3}}, \qquad z = \frac{c}{\sqrt{3}}.$$

These values (5.13) furnish the desired dimensions of the box, given an ellipsoid with semiaxes a, b, and c.

Now let us return to the original problem. We have seen that for given values a, b, c, the values (5.13) for x, y, z give the lengths of the half sides of a rectangular box of maximum volume. But in the other direction, is it true that for given values x, y, z, the values

(5.14) $$a = \sqrt{3}\,x, \qquad b = \sqrt{3}\,y, \qquad c = \sqrt{3}\,z$$

give the minimum volume for a containing ellipsoid? Whether the executors were perspicacious or just lucky, they were right in following their mathematical hunch. Here is a proof of that fact:

The volume W of the ellipsoid (5.10) is given by the formula

$$W = \frac{4}{3}\pi\,abc,$$

and for given positive x, y, z we want to choose positive a, b, c satisfying eq. (5.10) in such a way as to minimize W.

In order to see the reciprocal relationship between the two problems, it is convenient to consider reciprocals; and since it is now x, y, z that are given and a, b, c to be determined, we also reverse ends of the alphabet and set

(5.15) $a = \dfrac{1}{X}, \quad b = \dfrac{1}{Y}, \quad c = \dfrac{1}{Z}; \qquad x = \dfrac{1}{A}, \quad y = \dfrac{1}{B}, \quad z = \dfrac{1}{C}.$

Eq. (5.10) then becomes

$$\frac{(1/A)^2}{(1/X)^2} + \frac{(1/B)^2}{(1/Y)^2} + \frac{(1/C)^2}{(1/Z)^2} = 1$$

or

(5.10′) $$\frac{X^2}{A^2} + \frac{Y^2}{B^2} + \frac{Z^2}{C^2} = 1;$$

and, subject to this constraint, we wish to choose X, Y, Z so as to minimize

$$W = \frac{4}{3}\pi\, abc.$$

Since the coefficient $4\pi/3$ is a constant, minimizing W is equivalent to minimizing

$$abc = \frac{1}{XYZ}$$

which, in turn, is equivalent to maximizing

(5.11′) $V' = 8XYZ.$

But this is just the problem we have already solved, namely, the problem of determining the box of maximum volume that can be inscribed in a given ellipsoid. It does not matter that *this* ellipsoid (5.10′) and box of volume (5.11′) have no physical existence relative to our football player; indeed, this fact serves to emphasize and dramatize the importance of pure mathematical analysis. The solution values [cf. eq. (5.13)] are

(5.13′) $X = \dfrac{A}{\sqrt{3}}, \qquad Y = \dfrac{B}{\sqrt{3}}, \qquad Z = \dfrac{C}{\sqrt{3}}.$

Substituting from (5.15) into (5.13′), we obtain (5.13) and (5.14).

Considering, in a typical case, a length $2x$ of 6 feet, a width $2y$ of 2 feet, and a height $2z$ of 1 foot, we obtain the values

$$a = 3\sqrt{3}, \qquad b = \sqrt{3}, \qquad c = \sqrt{3}/2,$$

for an ellipsoid of minimum volume that would contain the box.

Because of the curvature of the elliptical vault, there would be ample room for the inclusion of a few well-worn footballs to accompany our gridiron hero to his place of rest.

5.9 Tangents

Let us now apply the theory of inequalities to the problem of finding a tangent to a given curve. Of course, we can do this only in an intuitive fashion at this time, since any precise definition of what we mean by a tangent to a curve at a point lies outside our chosen domain of discourse.

Consider a curve determined by an equation of the form $y = f(x)$, as shown in Fig. 5.10, and a line, with equation

(5.16) $mx + ny = k,$

that cuts the curve in the points P and Q. Let us move the line parallel to itself [this merely involves changing the value of k in the equation (5.16) of our line] until these two points become coincident at R, as shown in Fig. 5.11. The line $mx + ny = k'$ may be called the *tangent line* to the curve $y = f(x)$ at the point R. Let us again point out that we are proceeding in an intuitive fashion without attempting to make the notion of a tangent precise. But you are familiar at least with tangents to circles and can see that the procedure does lead to the desired result in this special case.

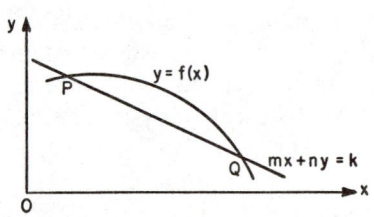

Figure 5.10. A curve and a cutting line

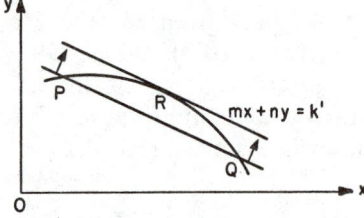

Figure 5.11. A curve and a tangent line

We shall use the foregoing procedure and the theory of inequalities in order to determine the tangent lines in a given direction to an ellipse. Suppose the ellipse (see Fig. 5.12) is given by the equation

$$\frac{x^2}{a^2} + \frac{y^2}{b^2} = 1,$$

and let (x_1, y_1) be one of the two points of tangency for lines of the form $mx + ny = k$, where m and n are fixed, $m^2 + n^2 \neq 0$, and k varies through all values. Observe that this point of tangency can be characterized in the following way:

(a) The point (x_1, y_1) lies on the ellipse; i.e., the values x_1, y_1 satisfy the equation

(5.17) $\frac{x_1^2}{a^2} + \frac{y_1^2}{b^2} = 1.$

(b) The point (x_1, y_1) lies on the line; i.e., the values x_1, y_1 satisfy the equation

$$mx_1 + ny_1 = k.$$

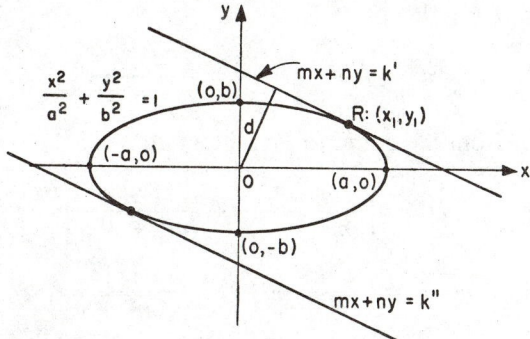

Figure 5.12. An ellipse and two parallel tangent lines

(c) The distance from the origin to the line $mx + ny = k$ is a *maximum* for all points (x_1, y_1) satisfying conditions (a) and (b), as k varies.

The distance d from the origin to the line $mx + ny = k$ is given by the formula

(5.18)
$$d = \frac{|k|}{\sqrt{m^2 + n^2}} = \frac{|mx_1 + ny_1|}{\sqrt{m^2 + n^2}}.$$

To see this, observe that the line

(5.19)
$$mx + ny = k$$

has slope $-m/n$ (see Fig. 5.13) and that a perpendicular line OP through the origin must therefore have the equation

(5.20)
$$y = \frac{n}{m}x.$$

Solving the system of linear equations (5.19) and (5.20), we find the

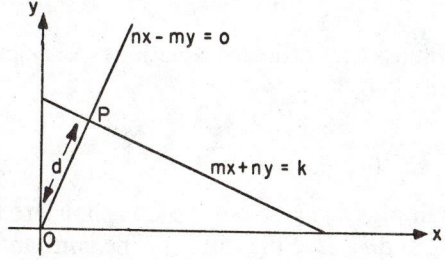

Figure 5.13. The distance formula

coordinates of P to be

$$x = \frac{km}{m^2 + n^2}, \qquad y = \frac{kn}{m^2 + n^2}.$$

The distance from the origin to P is therefore

$$d = OP = \left[\frac{k^2 m^2}{(m^2 + n^2)^2} + \frac{k^2 n^2}{(m^2 + n^2)^2} \right]^{1/2}$$

$$= \left(\frac{k^2}{m^2 + n^2} \right)^{1/2} = \frac{|k|}{\sqrt{m^2 + n^2}},$$

and, since $k = mx_1 + ny_1$ by condition (b), formula (5.18) is proved.

It follows that the problem of determining the tangent line is that of maximizing the expression (5.18) over all x_1 and y_1, subject to the constraint (5.17).

An application of Cauchy's inequality, i.e., the inequality (4.38) in Sec. 4.4(a) of Chapter 4, yields

$$d = \frac{|mx_1 + ny_1|}{\sqrt{m^2 + n^2}} = \frac{|(am)(x_1/a) + (bn)(y_1/b)|}{\sqrt{m^2 + n^2}}$$

(5.21)
$$\leq \left(\frac{a^2 m^2 + b^2 n^2}{m^2 + n^2} \right)^{1/2} \left(\frac{x_1^2}{a^2} + \frac{y_1^2}{b^2} \right)^{1/2}$$

$$= \left(\frac{a^2 m^2 + b^2 n^2}{m^2 + n^2} \right)^{1/2}.$$

The points of tangency are determined by two conditions. Namely, they must lie on the ellipse, so that (x_1, y_1) satisfies (5.17); and the distance (5.18) must attain its maximum value, so that the equality sign holds in (5.21), which is the case if and only if

(5.22)
$$\frac{x_1/a}{am} = \frac{y_1/b}{bn}.$$

The solution of the system of linear equations (5.17) and (5.22) yields the values

(5.23) $$x_1 = \pm \frac{a^2 m}{(a^2 m^2 + b^2 n^2)^{1/2}}, \qquad y_1 = \pm \frac{b^2 n}{(a^2 m^2 + b^2 n^2)^{1/2}},$$

where either both plus signs or both minus signs are taken, and the desired values k' and k'' (see Fig. 5.12) of the constant k can then be found by substitution in eq. (5.19).

5.10 Tangents (Concluded)

With a slight amount of additional effort, we can obtain a much more elegant result, resolving the problem of finding the tangent to an ellipse at a *given point* on the ellipse, rather than the tangent in a given direction.

In place of solving for x_1 and y_1 in terms of m and n, we need only to solve for m and n in terms of x_1 and y_1. It is that simple! From eq. (5.22), we see that we must have

$$m = \frac{rx_1}{a^2}, \qquad n = \frac{ry_1}{b^2}$$

where r is some as-yet-undetermined constant of proportionality. Using these values in the equation $mx + ny = k$, we obtain the equation of the tangent line in the form

$$\left(\frac{rx_1}{a^2}\right)x + \left(\frac{ry_1}{b^2}\right)y = k$$

or

$$\frac{xx_1}{a^2} + \frac{yy_1}{b^2} = \frac{k}{r}.$$

Since, on the one hand, (x_1, y_1) is a point on this tangent line, and since, on the other hand, it is also on the ellipse—that is, (x_1, y_1) satisfies equation (5.17)—we must have $k/r = 1$. Hence, we obtain the very simple and elegant result that the tangent to the ellipse

$$\frac{x^2}{a^2} + \frac{y^2}{b^2} = 1$$

at the point (x_1, y_1) on the ellipse has the form

$$\frac{xx_1}{a^2} + \frac{yy_1}{b^2} = 1.$$

As you will see when you study calculus, this is the same result that one obtains through the use of derivatives.

Exercises

1. Determine which of the points $(5, -3)$, $(3, 5)$, and $(7, 0)$ is on the ellipse

$$\frac{x^2}{50} + \frac{y^2}{18} = 1.$$

2. For $x = 2$, determine the values y such that the point (x, y) lies on the ellipse in Exercise 1. Also for $x = 2$, determine the values of y such that the point (x, y) lies either inside or on the ellipse.

3. Determine the equation of the tangent to the ellipse in Exercise 1, at the point $(-5, 3)$ on the ellipse.

4. Solve the system of equations

$$\frac{x^2}{a^2} + \frac{y^2}{b^2} = 1,$$

$$mx + ny = k,$$

where a, b, m, and n are fixed, choosing k so that the resulting quadratic equation has a double root. (You will find that $k^2 = a^2m^2 + b^2n^2$.) Compare your answer with the values of x_1, y_1 given at the end of Sec. 5.9, page 96.

CHAPTER SIX

Properties of Distance

6.1 Euclidean Distance

The distance with which you are familiar, between two points $P:(x_1, y_1)$ and $Q:(x_2, y_2)$ in the (x, y)-plane, is called *Euclidean* distance. We shall denote it by $d_2(PQ)$; it is given by the formula

$$(6.1) \qquad d_2(PQ) = [(x_2 - x_1)^2 + (y_2 - y_1)^2]^{1/2}.$$

We shall now enumerate some of the properties that characterize this distance function.

1. The distance between two points depends only on the position of one relative to the other; that is, it depends only on the differences $x_2 - x_1$ and $y_2 - y_1$ of their coordinates. This property (that the distance between two points does not change when both points are shifted by an equal amount in the same direction) is called *translation invariance.*

2. The distance from a point P to a point Q is equal to the distance from Q to P. This is seen by verifying, in (6.1), that

$$d_2(PQ) = d_2(QP).$$

Property 2 is usually called the *symmetry* of the distance function.

3. The *triangle inequality*

$$d_2(PR) \leq d_2(PQ) + d_2(QR)$$

is satisfied by the distance function (6.1); see Sec. 4.6.

4. The distance $d_2(PQ)$ between any two points is nonnegative for all P' and Q. That is,

$$d_2(PQ) \geq 0;$$

the sign of equality holds if and only if the points P and Q coincide. This is often called the *positivity* of the distance function; it follows immediately from the definition (6.1).

5. If P has coordinates (x, y) and Q has coordinates (ax, ay), where a is a nonnegative constant, then

$$d_2(OQ) = ad_2(OP);$$

here O denotes the origin $(0, 0)$. This property is sometimes called the *homogeneity* of the distance function, and it holds because

$$d_2(OQ) = [(ax)^2 + (ay)^2]^{1/2} = [a^2(x^2 + y^2)]^{1/2}$$
$$= a(x^2 + y^2)^{1/2} = a\, d_2(OP).$$

The Euclidean distance has still another property:

6. The Euclidean distance between two points remains unchanged if the (x, y)-plane is rotated about the origin through some angle. This property is sometimes called *rotation invariance*.

6.2 City-Block Distance

It turns out that many other useful and interesting "non-Euclidean" distances can be defined. In order to be called a "distance," a function of P and Q must have the properties 1 through 5 that we just verified for the familiar distance (6.1). The Euclidean distance d_2 alone has all six properties.

As an example, let us invent a "city-block distance" by determining the actual length of a path between two addresses $P:(x_1, y_1)$ and $Q:(x_2, y_2)$ in your home town, assuming that all the streets are laid out strictly north–south and east–west, and that there are no empty lots to cut across; see Fig. 6.1. Any path from P to Q is made up exclusively of horizontal and vertical pieces, so that the distance $d(PQ)$ we would have to traverse consists of the sum of all the horizontal distances and all the vertical distances making up the path from P to

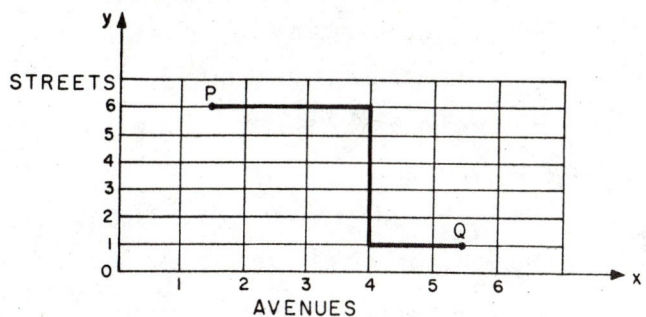

Figure 6.1. City-block distance

Q. We shall therefore define *city-block distance* d_1 by

$$(6.2) \qquad d_1(PQ) = |x_1 - x_2| + |y_1 - y_2|,$$

although, strictly speaking, this does not quite accurately describe the situation. It is not correct for the case shown in Fig. 6.2, where the addresses P and Q lie between the same two north–south (or east–west) streets; in this case, the traveler is forced to reverse directions in the course of his walk. Nevertheless, let us take (6.2) as the definition of a new "non-Euclidean" distance. After all, our city-block example just served as a motivation. (If our blocks are very small, (6.2) is fairly accurate. More precisely, our new distance function is the minimum distance required in traveling from P to Q when constrained to move only in the four principal compass directions.)

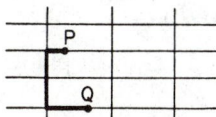

Figure 6.2. City-block distance, exceptional case

Next, let us see if d_1, as defined by (6.2), has the five properties required of a distance.

Since only the difference of the coordinates enters the expression (6.2), our new distance is certainly translation invariant, so that it has property 1.

Since $d_1(PQ) = d_1(QP)$, d_1 is symmetric, and thus it has property 2.

To establish the triangle inequality

$$d_1(PR) \leq d_1(PQ) + d_1(QR),$$

let P, Q, R have coordinates (x_1, y_1), (x_2, y_2), and (x_3, y_3), and write

$$d_1(PR) = |x_1 - x_3| + |y_1 - y_3|$$
$$= |x_1 - x_2 + x_2 - x_3| + |y_1 - y_2 + y_2 - y_3|;$$

since, by Theorem 3.2, page 42,

$$|x_1 - x_2 + x_2 - x_3| \leq |x_1 - x_2| + |x_2 - x_3|,$$
$$|y_1 - y_2 + y_2 - y_3| \leq |y_1 - y_2| + |y_2 - y_3|,$$

we have

$$d_1(PR) \leq |x_1 - x_2| + |y_1 - y_2| + |x_2 - x_3| + |y_2 - y_3|$$
$$= d_1(PQ) + d_1(QR).$$

The city-block distance certainly satisfies our fourth condition since the absolute value of any real number is always nonnegative. It is positive unless P and Q coincide.

Property 5 is easily verified since, for $a \geq 0$,

$$|ax| + |ay| = a[|x| + |y|].$$

Next, let us try to generalize the notion of a circle from Euclidean to city-block geometry. In Euclidean geometry, a circle is the locus of points equidistant from a fixed point. Let us carry this definition over to our new geometry. According to (6.2), the "unit circle" with center at the origin $O:(0, 0)$ would be given by the equation

$$d_1(OP) = |x| + |y| = 1.$$

Its graph in the ordinary Euclidean plane is shown in Fig. 6.3 (see also Fig. 3.14).

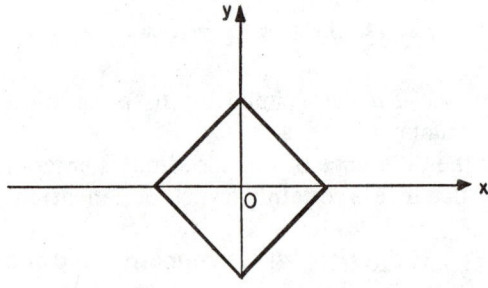

Figure 6.3. The unit circle in city-block geometry

6.3 Some Other Non-Euclidean Distances

Suppose that we now define the "distance" between the origin O and an arbitrary point P to be given by the expression

$$(6.3) \qquad d_p(OP) = (|x|^p + |y|^p)^{1/p}$$

for some fixed value of $p \geq 1$, and generally, that we define the distance between any two points $P:(x_1, y_1)$ and $Q:(x_2, y_2)$ to be given by

$$(6.4) \qquad d_p(PQ) = [|x_1 - x_2|^p + |y_1 - y_2|^p]^{1/p}.$$

Again, we check the five distance properties. The distance (6.4) certainly is translation invariant. Moreover, it is symmetric; that is, $d_p(PQ) = d_p(QP)$. Thirdly, the triangle inequality follows from formula (4.53′), which was proved at the end of Sec. 4.8 from the Minkowski inequality. Property 4 on positivity is also satisfied by (6.4); and finally, for $P:(x, y)$ and $Q:(ax, ay)$, we have

$$d_p(OQ) = [|ax|^p + |ay|^p]^{1/p} = (a^p)^{1/p}[|x|^p + |y|^p]^{1/p} = a\, d_p(OP),$$

so that our distance d_p also possesses the fifth property.

In this case the "unit circle," i.e., the locus of points at a distance 1 from the origin, is given by

$$|x|^p + |y|^p = 1.$$

Just what such locus curves look like depends on the particular value of p. For example, when $p = 1$, we are back at the city-block unit

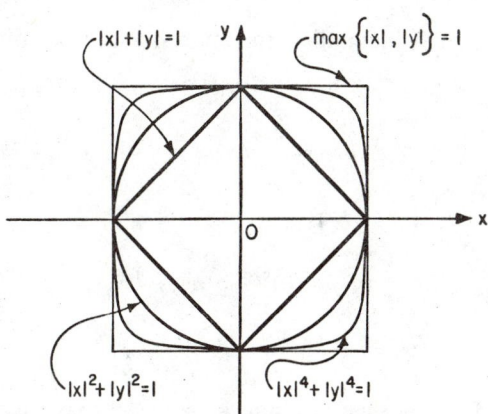

Figure 6.4. Euclidean graphs of non-Euclidean "unit circles"

circle; for $p = 2$, we have the usual Euclidean unit circle. For $p = 1, 2, 4$, we have the inequalities

$$|x| + |y| \geq (|x|^2 + |y|^2)^{1/2} \geq (|x|^4 + |y|^4)^{1/4},$$

which are easily verified by squaring. The Euclidean graphs of the corresponding "unit circles" are shown in Fig. 6.4; the curve for $p = 1$ is contained in that for $p = 4$. Is it true, in general, that the "unit circles" for distances defined by (6.3) are situated in such a way that the one for any fixed p contains those corresponding to smaller p? If so, do these curves become larger and larger as p increases?

To answer the first question, we first formulate an equivalent question: Does the inequality

$$[|x|^n + |y|^n]^{1/n} \geq [|x|^m + |y|^m]^{1/m}$$

hold whenever $m \geq n \geq 1$? It does indeed. Let us give a proof for $m = 3$, $n = 2$, and leave the general case for the reader as an exercise.

In order to avoid writing so many absolute-value signs, let us introduce

$$a = |x|, \qquad b = |y|.$$

We must then prove

$$(a^2 + b^2)^{1/2} \geq (a^3 + b^3)^{1/3} \qquad \text{for } a, b \geq 0.$$

We write

$$a^3 + b^3 = aa^2 + bb^2$$

and apply Cauchy's inequality (in square-root form), obtaining

$$(6.5) \qquad (a^2 + b^2)^{1/2}(a^4 + b^4)^{1/2} \geq a^3 + b^3.$$

Since

$$a^2 + b^2 \geq (a^4 + b^4)^{1/2},$$

(6.5) yields

$$(a^2 + b^2)^{1/2}(a^2 + b^2) \geq a^3 + b^3$$

or

$$(a^2 + b^2)^{3/2} \geq a^3 + b^3,$$

whence

$$(a^2 + b^2)^{1/2} \geq (a^3 + b^3)^{1/3}.$$

The proof for arbitrary rational $m \geq n \geq 1$ can be obtained by a corresponding application of Hölder's inequality. It is not difficult to see from the above discussion that, as p approaches 1, the "unit circles"

$$|x|^p + |y|^p = 1, \quad p > 1,$$

approach the square $|x| + |y| = 1$ shown in Fig. 6.3.

In order to see what happens as p gets larger and larger, we observe that

$$(6.6) \quad \max\{|x|^p, |y|^p\} \leq |x|^p + |y|^p \leq 2 \max\{|x|^p, |y|^p\}.$$

Since

$$[\max\{|x|^p, |y|^p\}]^{1/p} = \max\{|x|, |y|\},$$

(6.6) yields

$$(6.7) \quad \max\{|x|, |y|\} \leq [|x|^p + |y|^p]^{1/p} \leq 2^{1/p} \max\{|x|, |y|\}.$$

Now consider what happens to the right-hand member of (6.7) as p becomes very large; p occurs only in the exponent $1/p$ of 2. As p becomes very large, $1/p$ becomes very small and $2^{1/p}$ therefore approaches $2^0 = 1$. In other words, (6.7) tells us that

$$d_p(OP) = [|x|^p + |y|^p]^{1/p}$$

approaches the distance

$$(6.8) \quad d_\infty(OP) = \max\{|x|, |y|\}$$

as p "approaches infinity." It can be shown that d_∞ has all five distance properties.

Now, what does the "unit circle"

$$\max\{|x|, |y|\} = 1$$

look like? It is given by the square with sides

$$(6.9) \quad \begin{aligned} |x| &= 1, \quad 0 \leq |y| \leq 1 \\ |y| &= 1, \quad 0 \leq |x| \leq 1. \end{aligned}$$

So we see that

$$d_p(PQ) = [|x_1 - x_2|^p + |y_1 - y_2|^p]^{1/p}$$

can be used as a distance function for all $p \geq 1$ and that, as p grows larger, the "unit circles"

$$|x|^p + |y|^p = 1$$

grow larger, approaching the square (6.9) as p approaches infinity;

in spite of the fact that these "unit circles" grow bigger and bigger, they never get out of the square (see Fig. 6.4).

The "unit circle," in all these cases, divides the (x, y)-plane into two regions: the *interior* of the "unit circle," comprised of all points at distance less than 1 from the origin, and the *exterior*, comprised of all points at distance greater than 1 from the origin. The set of points defined by the inequality

$$d(OP) \leq 1$$

is sometimes called the *unit disc,* and the "unit circle"

$$d(OP) = 1$$

is called the *boundary* of the unit disc.

Some general remarks are in order. The Euclidean distance, as pointed out earlier, is invariant under translations (property 1) and rotations (property 6), i.e., under the so-called *displacements* or *rigid-body motions.* The other distances treated above also do not change under translation, but they do change under rotation. In fact, we can see from Fig. 6.4 that the city-block distance d_1 goes into the distance $d_\infty = \max\{|x_2 - x_1|, |y_2 - y_1|\}$ (stretched by a factor $\sqrt{2}$) when the (x, y)-plane is rotated through $45°$ about the point (x_1, y_1). It can be shown (but we shall not do so here) that the Euclidean distance is completely characterized by the six properties enumerated in Sec. 6.1; this means that the only distance that has, in addition to the five required properties, the property of rotation invariance is the Euclidean distance.

6.4 Unit Discs

There are many other functions having the five properties required of a distance; we have considered only a few.

We might ask the following question: Given a set S of points (which we shall henceforth call a *point set*) in the (x, y)-plane, such that S contains the origin in its interior, under what conditions does it represent the unit disc belonging to some distance d? In other words, under what conditions does there exist a distance function d for which S contains exactly those points P that are characterized by the inequality

$$d(OP) \leq 1?$$

We claim that, *for a given point set S having the origin in its interior there exists a distance function d such that S plus its boundary is the unit disc for d if and only if the following conditions are satisfied:*

(a) *The point set S is symmetric with respect to the origin.*

(b) *The point set S is convex.*

A point set S is symmetric with respect to the origin if, for each point (x, y) belonging to S, the point $(-x, -y)$ also belongs to S. It is convex if, for each pair of points in S, the line segment joining these points is entirely in S. See Fig. 6.5, (a) and (b).

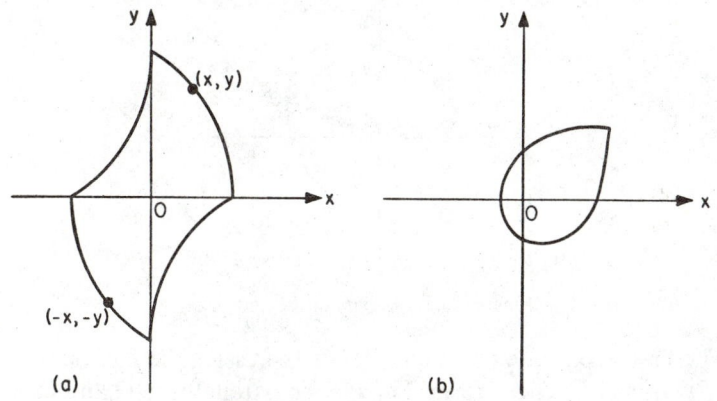

Figure 6.5. Point sets in the plane
(a) Point set symmetric with respect to the origin
(b) Convex Point set

We shall first prove that property (a) holds if d is a distance function: *If d is a distance function and S its unit disc, then S is symmetric with respect to the origin.* In other words, if d is a distance function and $P:(x, y)$ satisfies $d(OP) \leq 1$, then the point $Q:(-x, -y)$ satisfies $d(OQ) \leq 1$.

If d is a distance, then it is translation invariant; hence, if we shift the points Q and O by an amount x in the horizontal direction and an amount y in the vertical, their distance $d(QO)$ will remain unchanged. But such a shift brings the point Q into the origin O and the origin O into the point P. Hence

$$d(QO) = d(OP) \leq 1,$$

and, since d is also symmetric,

$$d(OQ) = d(QO) \leq 1.$$

Accordingly Q lies in the unit disc.

Next, let us prove that if d is a distance function then property (b) holds by showing that, *if P and Q are any two points in the unit disc, then the line segment PQ lies entirely in the unit disc.* Symbolically, the problem is expressed thus: Given any two points P, Q such that $d(OP) \leq 1$, $d(OQ) \leq 1$ for some distance function d, and any given point R on the segment PQ, prove that $d(OR) \leq 1$. This is immediately seen to be true if $P = Q$ or if either P or Q is at the origin, so we may assume that O, P, and Q are distinct points.

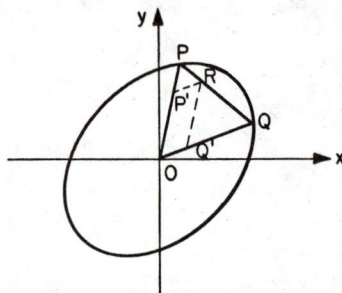

Figure 6.6. Convexity of the unit disc

The first step in the proof consists in expressing the fact that R lies on PQ in a convenient form. Let $P'R$ be parallel to OQ and let $Q'R$ be parallel to OP (see Fig. 6.6). Then we claim that

$$(6.10) \qquad d(OP') = a\, d(OP), \qquad d(OQ') = b\, d(OQ),$$

where $a > 0$, $b > 0$, and $a + b = 1$. To see this, we shall first employ the Euclidean theory of proportions (similar triangles). Using the same symbol AB for a Euclidean line segment and its length, we have

$$(6.11) \qquad a = \frac{OP'}{OP} = \frac{QR}{QP}, \qquad b = \frac{OQ'}{OQ} = \frac{PR}{PQ},$$

where a and b simply denote these ratios and hence are positive numbers. Moreover, by adding these ratios, we obtain

$$a + b = \frac{QR + RP}{QP} = \frac{QP}{QP} = 1.$$

Since P and P' lie along the same straight line through O, their coordinates are proportional. We may therefore make use of property 5 of the Euclidean distance and deduce, from (6.11), that if P has coordinates (x_1, y_1), then P' has coordinates (ax_1, ay_1). Similarly,

ACQUISITIONS/SERIALS
PROCESS SLIP

LOCATION:

RUSH _____

ALUMNI _____   RESERVE _____

CRC _____   SCI _____

MEDIA _____   STACKS ✓

REF _____   TR _____

Other:

S/I _RK_

STATUS:

ADD _____   CALL NO.

SO _____

PER _____

since Q and Q' are on the same line through O, if Q has coordinates (x_2, y_2), then Q' has coordinates (bx_2, by_2). We now use the fact that our distance d also has property 5, so that

$$(6.12) \qquad d(OP') = a\, d(OP), \qquad d(OQ') = b\, d(OQ).$$

Finally, we use property 3 (the triangle inequality; see Fig. 6.6):

$$d(OR) \leq d(OP') + d(P'R),$$

or, since $d(P'R) = d(OQ')$ by property 1,

$$d(OR) \leq d(OP') + d(OQ').$$

Substituting from (6.12), we have

$$d(OR) \leq a\, d(OP) + b\, d(OQ),$$

and, since $d(OP) \leq 1$, $d(OQ) \leq 1$, and $a + b = 1$,

$$d(OR) \leq a + b = 1,$$

which, by definition, means that R is in the unit disc.

So far we have shown that, if d is a distance and S its unit disc, then S is symmetric with respect to the origin and convex. We must still prove the converse: *If S is a point set containing the origin in its interior, and S is convex and symmetric with respect to the origin, then there exists a distance function d for which S is the unit disc.*

We shall indicate how such a distance d may be defined, but we shall leave to the reader the job of verifying that d has the five prescribed distance properties.

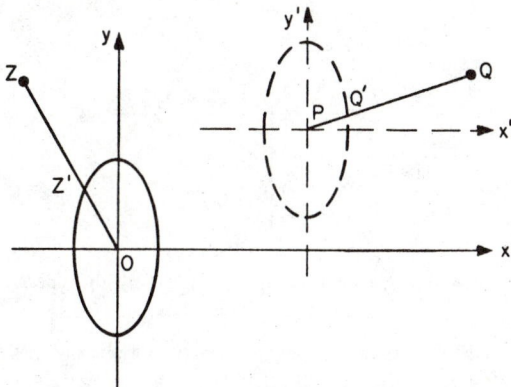

Figure 6.7. $d(OZ) = \dfrac{OZ}{OZ'}, \qquad d(PQ) = \dfrac{PQ}{PQ'}$

Let S be a point set of the prescribed sort; see Fig. 6.7. Let Z be any point other than O in the plane. Draw the ray from O through Z and let it intersect the boundary of S in the point Z'. (It follows from the convexity of S that there is just one such intersection of the ray with the boundary of S.) Then calculate the ratio

$$r = \frac{OZ}{OZ'}$$

of the Euclidean distances OZ to OZ' and define the distance from O to Z to be this ratio; i.e., let

$$d(OZ) = r.$$

Observe that $d(OZ)$ is less than 1, equal to 1, or greater than 1 depending on whether Z is an interior point of S, a boundary point of S, or an exterior point of S, respectively.

To define $d(PQ)$ for any points P and Q, shift the coordinates as indicated in Fig. 6.7 and proceed as before.

In an ordinary Euclidean circle, the ratio of the circumference to the diameter is denoted by the symbol π and is approximately 3.14. In the exercises below, the task is to find the ratio r of the non-Euclidean length of the circumference to that of the diameter of some non-Euclidean unit circles. We shall analyze the situation in one particular case, below, and leave other simple cases for the reader to investigate.

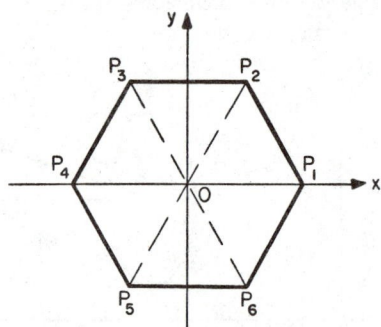

Figure 6.8. A regular hexagon symmetric with respect
to the co-ordinate axes

EXAMPLE. Let S consist of the points of a regular hexagon symmetrically situated with respect to the coordinate axes, together with the points of its interior; see Fig. 6.8. Since S is convex and symmetric with respect to the origin, it may be considered the unit disc for some distance function d. Since the non-Euclidean radius of the *unit* disc is 1 by definition, its diameter is 2. To calculate the circumference, we observe that, because of the translation invariance of d, the following relations hold:

$$d(P_1P_2) = d(OP_3) = 1, \quad d(P_2P_3) = d(OP_4) = 1, \quad d(P_3P_4) = d(OP_5) = 1,$$
$$d(P_4P_5) = d(OP_6) = 1, \quad d(P_5P_6) = d(OP_1) = 1, \quad d(P_6P_1) = d(OP_2) = 1.$$

If we add the lengths of all these segments, we find that the circumference is 6. Hence, the desired ratio is

$$r = \frac{6}{2} = 3.$$

Exercises

Calculate the ratio of the non-Euclidean length of the circumference to that of the diameter for the following unit discs S:

1. S is the unit disc for the city-block distance $|x_2 - x_1| + |y_2 - y_1|$; see Fig. 6.3.

2. S is the unit disc for the distance function
$$d(OP) = \max \{|x|, |y|\}.$$

3. The unit disc S consists of the interior and boundary points of a regular octagon with center at the origin.

4. The unit disc S consists of the interior and boundary points of a regular ten-sided polygon with center at the origin.

6.5 Algebra and Geometry

What we have observed in the preceding sections is that geometric intuition can be used to derive interesting algebraic results. In two or three dimensions, this technique works well. As soon as we turn to a discussion of n-dimensional geometry, for $n \geq 4$, the situation reverses. Now, we often rely on algebra to make geometric definitions and to establish geometric results.

Let us briefly illustrate this idea. Consider a set of n real quantities $x_1, x_2, \ldots, x_n$ as constituting a point P in n-dimensional space. The Euclidean distance between the two points $P:(x_1, x_2, \ldots, x_n)$ and $Q:(y_1, y_2, \ldots, y_n)$ is *defined* to be

$$(6.13) \qquad d(PQ) = [(x_1 - y_1)^2 + (x_2 - y_2)^2 + \cdots + (x_n - y_n)^2]^{1/2}.$$

For $n = 2$, (6.13) reduces to the familiar expression for the distance between two points (x_1, x_2) and (y_1, y_2) in the plane. If we denote the origin $(0, 0, \ldots, 0)$ by O and the point $(x_1 + y_1, x_2 + y_2, \ldots, x_n + y_n)$

by R, then the n-dimensional triangle inequality

$$d(OP) + d(PR) \geq d(OR)$$

would read

$$(6.14) \quad [x_1{}^2 + x_2{}^2 + \cdots + x_n{}^2]^{1/2} + [y_1{}^2 + y_2{}^2 + \cdots + y_n{}^2]^{1/2}$$
$$\geq [(x_1 + y_1)^2 + (x_2 + y_2)^2 + \cdots + (x_n + y_n)^2]^{1/2}.$$

This is a valid inequality, as was noted on page 70 in Sec. 4.6.

Next, let us define the cosine of the angle θ between the lines OP and OQ by

$$(6.15) \quad \cos \theta = \frac{x_1 y_1 + x_2 y_2 + \cdots + x_n y_n}{(x_1{}^2 + x_2{}^2 + \cdots + x_n{}^2)^{1/2}(y_1{}^2 + y_2{}^2 + \cdots + y_n{}^2)^{1/2}}.$$

The Cauchy inequality [see Sec. 4.4(d), inequality (4.45)] shows that $|\cos \theta| \leq 1$.

We now have the foundations for an analytic geometry of n dimensions.

SYMBOLS

$\lvert a \rvert$	absolute value of a
N	set of all negative numbers
O	set of one element, namely, the number 0
P	set of all positive numbers
ε	is a member of
$\notin$	is not a member of
$=$	equal to
$\neq$	not equal to
$>$	greater than
$\not>$	not greater than
$<$	less than
$\not<$	not less than
$\geq$	greater than or equal to
$\not\geq$	neither greater than nor equal to
$\leq$	less than or equal to
$\not\leq$	neither less than nor equal to
$\not\gtrless$	neither greater than nor less than
$\sqrt{}$	nonnegative square root
$\{a_1, a_2, \ldots, a_n\}$	set of elements $a_1, a_2, \ldots, a_n$
$\max\{a_1, a_2, \ldots, a_n\}$	greatest element of $\{a_1, a_2, \ldots, a_n\}$
$\min\{a_1, a_2, \ldots, a_n\}$	least element of $\{a_1, a_2, \ldots, a_n\}$
$\{a_1, a_2, \ldots, a_n\}^+$	$\max\{0, a_1, a_2, \ldots, a_n\}$
$\{a_1, a_2, \ldots, a_n\}^-$	$\min\{0, a_1, a_2, \ldots, a_n\}$
$\operatorname{sgn} x$	0 for $x = 0$, 1 for $x > 0$, -1 for $x < 0$.

Answers to Exercises

Chapter 1
Pages 12–13

1.

$$-3 < -2 < -1.5 < -1 < 3 - \pi < 0 < \pi - 3 < \sqrt{2} < 2 < 3$$

2. (a) ε (b) $\notin$ (c) $\notin$ (d) $\notin$ (e) ε (f) $\notin$ (g) ε (h) $\notin$ (i) $\notin$ (j) $\notin$

3. (a) N (b) P (c) P (d) N (e) P (f) P (g) P (h) N (i) P (j) O

4. (a) $<$ (b) $>$ (c) $>$ (d) $<$ (e) $>$ (f) $>$
   (g) $>$ (h) $<$ (i) $>$ (j) $=$

5. (a) T (b) T (c) T (d) F (e) T (f) T (g) F (h) T (i) F (j) F

6. $2,\ \pi - 3,\ -(3 - \pi)^2,\ a/(c - b),\ 0,\ -\sqrt{b^2 - 4ac}$.

7. (a) $\geq$ (b) $\gtrless$ (c) $\leq$ (d) $>$ (e) $<$ (f) $=$

8. $p > 0,\ -n > 0,$ so $p - n > 0,$ so $p > n.$

9. $a = b$

10. True for $n = 1$ by hypothesis; induction step by Axiom II.

11. Numbers 1 and 2 were shown in text to be "positive," so $1 + 2 = 3$ is "positive" by Axiom II. Let $a = 1/3$; then $3a = 1$, so a is "positive" since 3 and 1 are "positive." Then $2a = 2/3$ is "positive" by Axiom II.

Chapter 2

Page 19

1. Add $a/2 < b/2$ first to $a/2 = a/2$, then to $b/2 = b/2$.

2. Both inequalities are equivalent to $(ad - bc)^2 \geq 0$.

3. Equivalent to $(a - b)^4 \geq 0$.

4. For $n = 2$, the theorem states: If $a_1 \geq a_2$, then $a_1 \geq a_2$, which is true. Suppose true for n and let $a_1 \geq a_2, \ldots, a_{n-1} \geq a_n, a_n \geq a_{n+1}$. Then $a_1 \geq a_n, a_n \geq a_{n+1}$, so $a_1 \geq a_{n+1}$. Sign of equality if and only if $a_1 = a_2, \ldots, a_{n-1} = a_n$, and $a_n = a_{n+1}$.

Pages 21–22

1. Intermediate step:

$$\frac{2}{\sqrt{ab}} \leq \frac{1}{a} + \frac{1}{b}.$$

Equality if and only if $a = b$.

2. In Exercise 1, let $b = 1/a$. Equality if and only if $a = 1$.

3. Add together $a^2 + b^2 \geq 2ab$, $b^2 + c^2 \geq 2bc$, $c^2 + a^2 \geq 2ca$, then divide by 2.

4. Equivalent to $a^2b^2(a - b)^2 \geq 0$.

5. Multiply $a^2 + b^2 \geq 2ab$ by c, $b^2 + c^2 \geq 2bc$ by a, $c^2 + a^2 \geq 2ac$ by b, then add.

6. Intermediate step:
$$(a^2 - b^2)^2 - (a - b)^4 = 4ab(a - b)^2.$$

7. Intermediate step:
$$(a^3 + b^3) - (a^2b + ab^2) = (a + b)(a - b)^2.$$

8. (3) $a = b = c$; (4) $a = b$ or at least one of them $= 0$;
   (5) $a = b = c$; (6) $a = b$ or at least one of them $= 0$; (7) $a = \pm b$.

Page 24

1. Equivalent to $(a - b)^2 \geq 0$. Sign of equality if and only if $a = b \geq 0$.

2. By Theorem 2.7 and the fact that c and d are of the same sign, $a^c - b^c$ and $a^d - b^d$ are of the same sign. Multiply out. Sign of equality if and only if $a = b$.

3. $a^2 + b^2 \geq 2ab$, $a + b \geq 2\sqrt{ab}$.

4. The equivalence of these inequalities can be proved by multiplying the first by $bd > 0$ and the second by $1/bd > 0$; see Theorem 2.3. The equivalence of the corresponding equations is proved in the same way.

5. Add $1 = 1$ to the inequality. Equality $a/b = c/d$ is equivalent to $ad = bc$.

6. Apply Theorem 2.7 with exponent -1, add $1 = 1$ to resulting inequality, then apply Theorem 2.7 again with exponent -1. Sign of equality as in Exercise 5.

7. Intermediate steps:

$$\frac{a + c}{b + d} - \frac{a}{b} = \frac{bc - ad}{b(b + d)} \geq 0, \quad \frac{c}{d} - \frac{a + c}{b + d} = \frac{bc - ad}{d(b + d)} \geq 0.$$

8. $30 < 33$; $22 < 25$; $18 < 21$; $42 < 45$.

9. If $a < b$ and $b < c$, then $a < c$.
If $a < b$ and $c < d$, then $a + c < b + d$. If $a < b$ and c is any real number, then $a + c < b + c$.
If $a < b$ and $c > 0$, then $ac < bc$. If $a < b$ and $c < 0$, then $ac > bc$.
If $a < b$ and $c < d$, then $a - d < b - c$. If $a < b$ and c is any real number, then $a - c < b - c$.
If $0 < a < b$ and $0 < c < d$, then $ac < bd$.
If $0 < a < b$ and $0 < c < d$, then $a/d < b/c$. In particular, for $a = b = 1$, if $0 < c < d$, then $1/d < 1/c$.
If $0 < a < b$, if m and n are positive integers, and if $a^{1/n}$ and $b^{1/n}$ denote positive nth roots, then

$$a^{m/n} < b^{m/n} \quad \text{and} \quad b^{-m/n} < a^{-m/n}.$$

Chapter 3

Pages 28–29

1. (a) -1 (b) π (c) 0 (d) 4 (e) 3 (f) 0 (g) π (h) 0 (i) 4 (j) 3

2. (a) -7 (b) $\sqrt{2}$ (c) -7 (d) 0 (e) -3 (f) -7 (g) 0 (h) -7 (i) 0 (j) -3

3. $(-1)(0) - (1)(0) = 0$.

4. Consider representative cases along with the definition of $\max \{ \quad \}$.

5. $\{a, b, c, d\} = \{-1, -1, -1, -1\}$.

6. One of $\{a, b\}^+$ and $\{c, d\}^+$ is equal to $\{a, b, c, d\}^+$; the other is nonnegative.

7. First and third inequalities: possibly an additional candidate, namely 0. Middle inequality: definition of max $\{\quad\}$ and min $\{\quad\}$. No; $a_1, a_2, \ldots, a_n$ would all have to be negative for first strict inequality, and all positive for third.

8. Multiply inequalities $a \geq b$, $a \geq c$ by -1 and apply Theorem 2.3.

9. $\{-a, -b\}^- = \min \{0, -a, -b\} = -\max \{0, a, b\} = -\{a, b\}^+$.

10. Consider representative cases along with definition of max $\{\quad\}$.

Pages 33–34

1. For each real number a, $-|a| \leq -a \leq |a|$. The first sign of equality holds if and only if $a \geq 0$, and the second if and only if $a \leq 0$.

2.

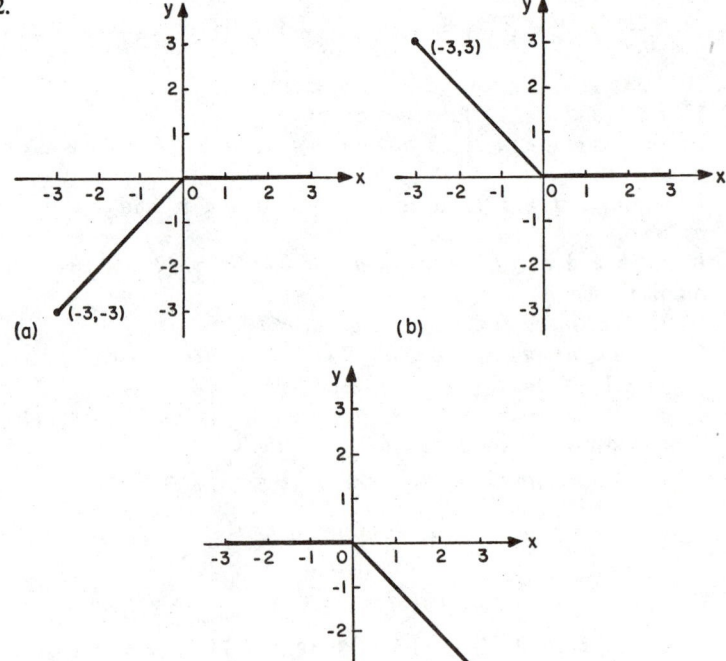

3. (a) is also the graph of (d), (g); (b) is also the graph of (e), (h); (c) is also the graph of (f), (i).

4.

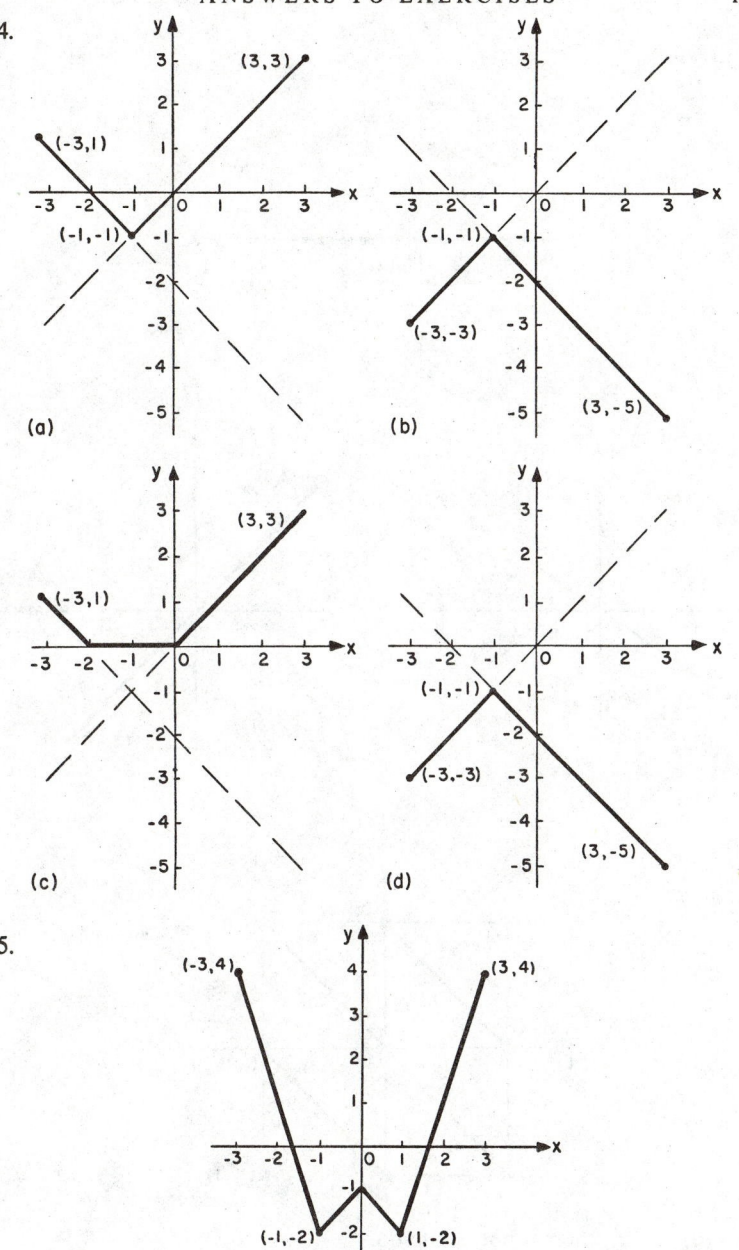

(a)

(b)

(c)

(d)

5.

6. Fig. 3.3, odd; Fig. 3.4, odd; Fig. 3.5, even; Fig. 3.6, even.

Page 36

1.

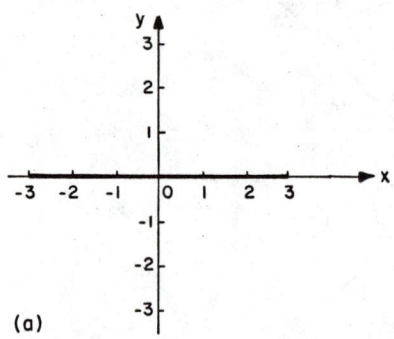

(a)

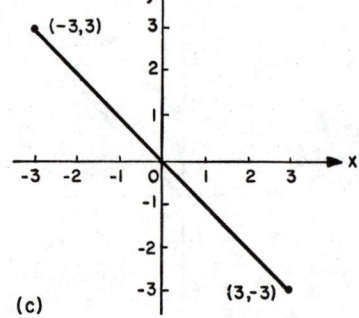

(b)

(c)

2.

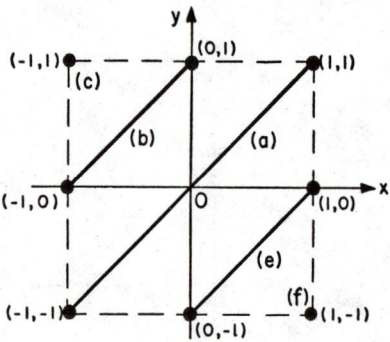

Part (d) is easy; there is nothing to do.

3.

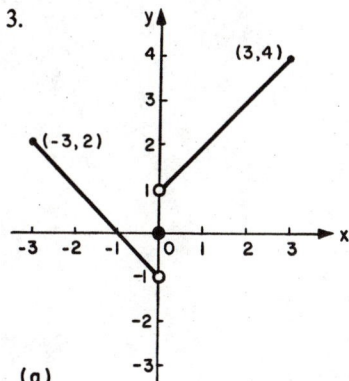

(a)

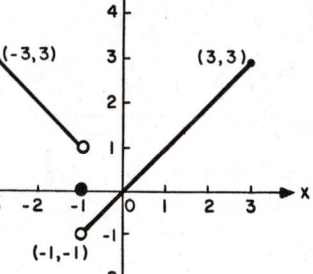

(b)

4.

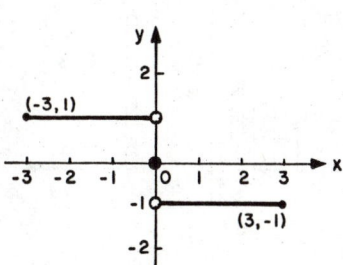

Page 39

1.

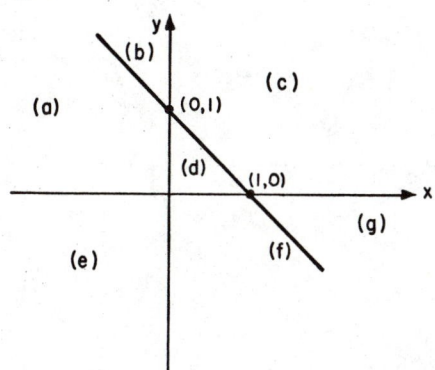

(a) $x \leq 0,\ y \geq 0,\ x+y \leq 1$;

(b) $x \leq 0,\ y \geq 0,\ x+y \geq 1$;

(c) $x \geq 0,\ y \geq 0,\ x+y \geq 1$;

(d) $x \geq 0,\ y \geq 0,\ x+y \leq 1$;

(e) $x \leq 0,\ y \leq 0,\ x+y \leq 1$;

(f) $x \geq 0,\ y \leq 0,\ x+y \leq 1$;

(g) $x \geq 0,\ y \leq 0,\ x+y \geq 1$.

2.

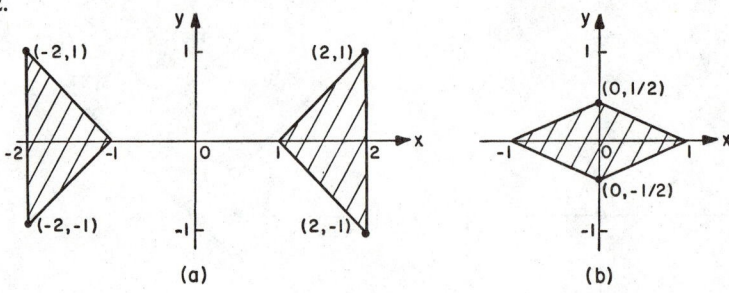

(a)   (b)

3.

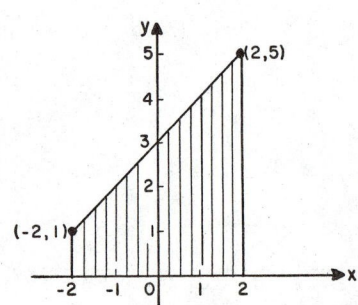

4.

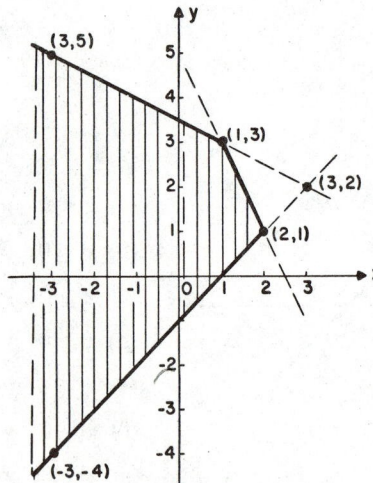

Incomplete graph.

5.

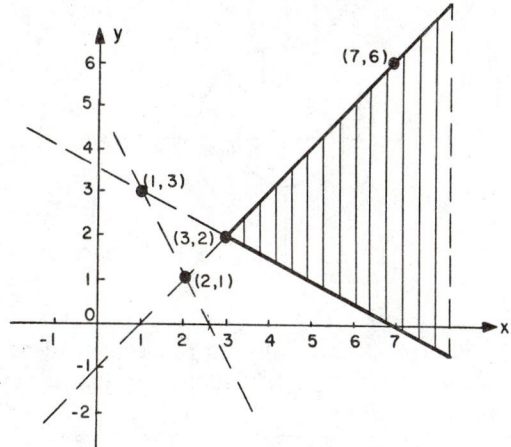

Incomplete graph.

Page 41

1.

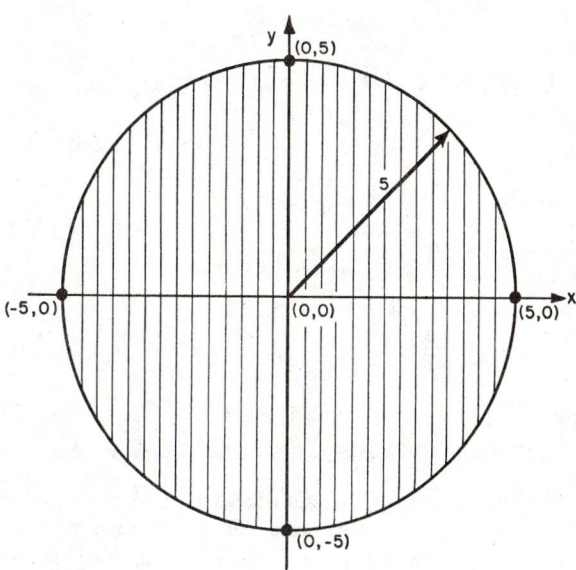

2.

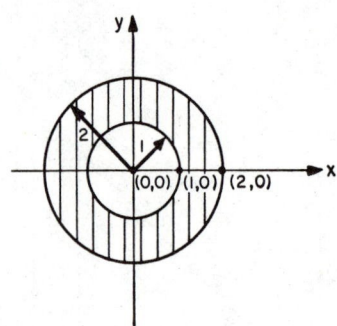

3. The y-axis, $x = 0$. Geometrically, the problem is to determine the locus of points equidistant from $(-1, 0)$ and $(1, 0)$. Analytically, it is to solve the equation

$$\sqrt{(x + 1)^2 + y^2} = \sqrt{(x - 1)^2 + y^2}.$$

Page 45

1. (a) $|-a| = \sqrt{(-a)^2} = \sqrt{a^2} \qquad = |a|$
   (b) $|ab| = \sqrt{(ab)^2} = \sqrt{a^2}\,\sqrt{b^2} = |a|\cdot|b|$
   (c) $\left|\dfrac{a}{b}\right| = \sqrt{\left(\dfrac{a}{b}\right)^2} = \dfrac{\sqrt{a^2}}{\sqrt{b^2}} \qquad = \dfrac{|a|}{|b|}.$

2. (a) $=$, (b) $>$, (c) $=$, (d) $=$, (e) $>$.

3. (a) $>$, (b) $=$, (c) $=$, (d) $=$, (e) $=$.

4. (a) $>$, (b) $=$, (c) $=$, (d) $>$, (e) $=$.

5. (a) $=$, (b) $>$, (c) $>$, (d) $=$, (e) $=$.

6. $\sqrt{(a - b)^2} \qquad \geq \sqrt{(\sqrt{a^2} - \sqrt{b^2})^2},$
   $(a - b)^2 \qquad \geq (\sqrt{a^2} - \sqrt{b^2})^2,$
   $a^2 - 2ab + b^2 \geq a^2 - 2\sqrt{a^2 b^2} + b^2,$
   $2\sqrt{a^2 b^2} \qquad \geq 2ab,$
   $|ab| \qquad \geq ab.$

7. Suppose $a^2 \leq b^2$ and $ab \geq 0$. Then
   $$\min\{a^2, b^2\} = a^2 = |a|\cdot|a| \leq |a|\cdot|b| = ab.$$

8. If $a \geq 0$, then $\sqrt{a^2} = a$; if $a < 0$, then $\sqrt{a^2} = -a > 0$. Similarly, $\sqrt{a^2}$ is 0 if $a = 0$, and otherwise it is the positive member of the set $\{a, -a\}$; $\sqrt{a^2} = \max\{a, -a\}$; $\sqrt{a^2} = \{a, -a\}^+$; the graph of $y = \sqrt{x^2}$ is shown in Fig. 3.6; and $\sqrt{a^2} = a\,\mathrm{sgn}\,a$.

Chapter 4
Page 49

1. (a) 4, 5; (b) 6, 7.5; (c) 6, 6.5; (d) 0, 10.

2. (a) $3p, 5p$; (b) $0, p/2$; (c) $2p, p^2 + 1$.

Pages 51–52

1. $a + b = \text{diameter} = 2r$, so $r = (a + b)/2$; by similar triangles, $a/h = h/b$, so $h = \sqrt{ab}$.

2. To show that the harmonic mean is less than or equal to the geometric mean, an intermediate step is
$$ab(a - b)^2 \geq 0;$$
or see Exercise 1 on page 21. The sign of equality holds if and only if $a = b$. To show that the harmonic mean is less than or equal to the arithmetic mean, you can now use the arithmetic-mean–geometric-mean inequality or can proceed directly, with intermediate step
$$(a - b)^2 \geq 0.$$

3. (a) 3.2, 4, 5; (b) 4.8, 6, 7.5; (c) $5.54^-, 6, 6.5$; (d) $5.83^+, 5.91^+, 6$; (e) 6, 6, 6.

4. Half the distance at each rate:
$$\frac{d}{2} = r_1 t_1 = r_2 t_2, \quad t_1 = \frac{d}{2r_1}, \quad t_2 = \frac{d}{2r_2}, \quad t = t_1 + t_2 = \frac{d}{2}\left(\frac{1}{r_1} + \frac{1}{r_2}\right),$$
$$d = rt = \frac{rd}{2}\left(\frac{1}{r_1} + \frac{1}{r_2}\right), \quad r = \frac{2}{(1/r_1) + (1/r_2)} = \frac{2r_1 r_2}{r_1 + r_2}.$$

Half the time at each rate:
$$d = rt = \frac{r_1 t}{2} + \frac{r_2 t}{2}, \quad r = \frac{r_1 + r_2}{2}.$$

By the result of Exercise 2, half the time at each rate would get you there sooner.

5. Let $b = 1/a$.

Pages 61–62

1. In the arithmetic-mean–geometric-mean inequality (4.19) set the first m_1 of the numbers a_i equal to the same value y_1, set the next m_2 of the numbers a_i equal to the same number y_2, and set the last m_k of the numbers a_i equal to the same value y_k, and observe that
$$m_1 + m_2 + \cdots + m_k = n.$$
This gives the first inequality. For the second inequality, set
$$\frac{m_1}{n} = r_1, \qquad \frac{m_2}{n} = r_2, \qquad \cdots \qquad , \qquad \frac{m_k}{n} = r_k.$$

2. $8.5, 9.1^+$; $0.5, 0.7^+$; p, p.

3. Intermediate step: $(a - b)^2 \geq 0$; or see Exercise 1 on page 24. Equality holds if and only if $a = b$. Since the root-mean-square is greater than or equal to the arithmetic mean, it is also greater than or equal to the geometric and harmonic means.

4. (a) A diagonal divides GH into two segments, one of length $a/2$, the other of length $b/2$.

(b) If $ABLK \sim KLDC$, then $AB/KL = KL/CD$ and $KL^2 = \sqrt{ab}$.

(c) The harmonic mean is

$$\frac{2ab}{a + b} = h.$$

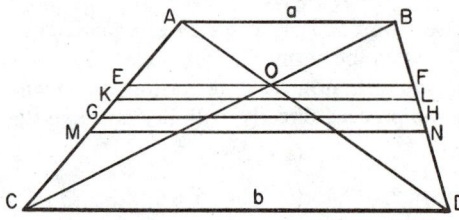

To show that $EF = h$, prove that $EO = OF$ and use similar triangles:

$$\frac{EO}{CD} = \frac{AE}{AC} = \frac{AC - EC}{AC} = 1 - \frac{EC}{AC}.$$

But

$$\frac{EC}{AC} = \frac{EO}{AB}.$$

Hence

$$\frac{EO}{CD} = 1 - \frac{EO}{AB} \quad \text{or} \quad EO\left(\frac{1}{AB} + \frac{1}{CD}\right) = 1;$$

thus

$$EO = \frac{AB \cdot CD}{AB + CD} = \frac{ab}{a + b} = \frac{1}{2}h$$

and

$$EF = 2EO = h.$$

(d) Set $MN = r$, let x and y be the altitudes of the constituent trapezoids so that $x + y$ is the altitude of $ABDC$. Then, by hypothesis,

$$\frac{r + a}{2} \cdot x = \frac{1}{2}\frac{a + b}{2}(x + y), \qquad \frac{r + b}{2} \cdot y = \frac{1}{2}\frac{a + b}{2}(x + y).$$

This system of simultaneous linear equations in x and y has a solution if and only if

$$r^2 = \frac{a^2 + b^2}{2};$$

hence r is the root-mean-square of a and b.

Chapter 5
Page 87

1. Add together $ab + bc + ca \le a^2 + b^2 + c^2$, from Exercise 3 on page 22, and $2(ab + bc + ca) = 2(ab + bc + ca)$, to get $3(ab + bc + ca) \le (a + b + c)^2$, whence

$$A = 2(ab + bc + ca) \le \frac{2}{3}(a + b + c)^2 = \frac{2}{3}\left(\frac{E}{4}\right)^2 = \frac{E^2}{24}.$$

Equality in Exercise 3 on page 22 if and only if $a = b = c$.

2. If the surface area A of a rectangular box is a fixed value A, then the sum of the lengths of the twelve edges is at least $2\sqrt{6A}$, and the box is a cube if and only if $E = 2\sqrt{6A}$. The proof is immediate from the inequality $A \le E^2/24$ of Exercise 1.

3. Let A = area, P = perimeter, a and b = lengths of sides, c = length of diagonal. By inequality between arithmetic mean and root-mean-square (Exercise 3 on page 62),

$$P = 2(a + b) = 4\left(\frac{a + b}{2}\right) \le 4\sqrt{\frac{a^2 + b^2}{2}} = 2\sqrt{2}c.$$

By inequality between geometric mean and arithmetic mean,

$$A = ab \le \left(\frac{a + b}{2}\right)^2 = \frac{P^2}{16}.$$

Hence

$$A \le \frac{c^2}{2}.$$

The signs of equality hold if and only if $a = b$. Hence the square has the greatest perimeter $(2\sqrt{2}c)$ and the greatest area $(c^2/2)$.

Pages 89–90

1. If $y = z$, then $s - y = s - z = x/2$, so
$$I^2 = s(s - x)(s - y)(s - z) = s(s - x)\frac{x}{2}\frac{x}{2} = \frac{s}{4}x^2(s - x).$$

If $x = y = z$, then $s - x = s - y = s - z = s/3$, so
$$E^2 = s(s - x)(s - y)(s - z) = \frac{s^4}{27}.$$

2. Substitute for E^2, I^2, and A^2 from Exercise 1, then multiply out and compare terms.

3. Each factor on the right is nonnegative, either because it is a square or because of its geometric meaning. Have $E^2 = I^2$ if and only if $x = 2s/3$, i.e., if and only if the isosceles triangle actually is equilateral. Have $I^2 = A^2$ if and only if either $x = s$ or $y = z$, i.e., if and only if the general triangle actually is isosceles.

4. The second inequality. See the discussion of $I^2 = A^2$ in the above solution of Exercise 3.

5. The first inequality. See the discussion of $E^2 = I^2$ in the above solution of Exercise 3.

6. The formula shows that the square of the area of the triangle is equal to the square of the area of the equilateral triangle having the same perimeter, diminished by a certain amount if the triangle is only isosceles, and diminished still more if the triangle is not even isosceles.

7. Let a and b = lengths of sides, c = length of hypotenuse, t = length of altitude. Then by similar triangles, $t = ab/c$. By the arithmetic-mean-geometric-mean inequality,

$$t = \frac{ab}{c} \leq \frac{a^2 + b^2}{2c} = \frac{c}{2},$$

with equality if and only if $a = b$, i.e., if and only if the right triangle is isosceles.

Pages 97–98

1. $(5, -3)$.

2. $\pm \frac{3}{5}\sqrt{46}$; $\quad -\frac{3}{5}\sqrt{46} \leq y \leq \frac{3}{5}\sqrt{46}$.

3. $\dfrac{-x}{10} + \dfrac{y}{6} = 1$.

4. If y is expressed in terms of x from the linear equation, and the resulting expression is substituted in the quadratic equation, then its discriminant is $4a^2b^2n^2(a^2m^2 + b^2n^2 - k^2)$; this expression vanishes if

$$k = \pm \sqrt{a^2m^2 + b^2n^2}.$$

The corresponding double roots are given on page 96, eqs. 5.23.

Chapter 6
Page 111

1.

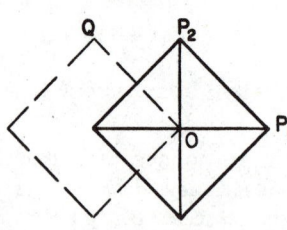

The city-block length of the diameter is 2 and that of each side is $d_1(P_1, P_2) = d_1(O, Q) = 2$. Hence the city-block length of the perimeter is $(4)(2) = 8$, and the desired ratio is

$$r = \frac{8}{2} = 4.$$

2.

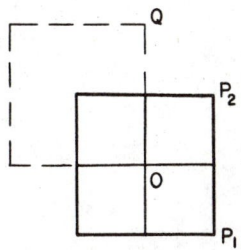

The non-Euclidean length of the diameter is 2 and that of each side is $d_\infty(P_1 P_2) = d_\infty(OQ) = 2$. Thus, the non-Euclidean length of the perimeter is 8 and

$$r = \frac{8}{2} = 4.$$

3. *Hint.* Each regular polygon with an even number of sides and with center at the origin can be placed in two different positions, each symmetric with respect to each coordinate axis. Try to prove that
(a) if the number N of sides of the polygon is divisible by 4, then the non-Euclidean length of each side is $2 \tan(180°/N)$ [and hence the perimeter has non-Euclidean length $2N \tan(180°/N)$],
(b) if the number N of sides is even but *not* divisible by 4, then the length of each side is $2 \sin(180°/N)$ [and hence the perimeter is $2N \sin(180°/N)$],
(c) the results (a) and (b) are valid for all possible positions of the polygon.
 Since $N = 8$ is divisible by 4, we have

$$\text{diameter} = 2,$$
$$\text{non-Euclidean length of perimeter} = 16 \tan(180°/8)$$
$$= 16 \tan 22.5°,$$
$$r = \frac{16 \tan 22.5°}{2} = 8(\sqrt{2} - 1)$$
$$\approx 3.314.$$

4. Since $N = 10$ is *not* divisible by 4, we have

$$\text{diameter} = 2,$$
$$\text{non-Euclidean length of perimeter} = 20 \sin(180°/10),$$
$$r = \frac{20 \sin 18°}{2} = 10 \sin 18° \approx 3.090.$$

Index

absolute value, 25–45
  algebraic characterization of, 40–41
  and classical inequalities, 73–75
  definition of, 26
  graph of, 30, 31
  and sign function, 34–36
addition of inequalities, 17
algebraic characterization of absolute value, 40–41
arithmetic mean, 48
arithmetic-geometric mean, 76–78
arithmetic–mean—geometric–mean inequality, 48–61, 73–74
axiom, 7

backward induction, 57–59
bound, upper, 12 (footnote)
Buniakowski, V., 63 (footnote)

Cauchy inequality, 62–67, 73–74
  geometric interpretation of, 63
Cauchy-Lagrange identity, 66

city-block distance, 100–102
classical inequalities, 73–75
complex numbers, 41
convex set, 107
cosine inequality, 65

Dido, problem of, 80–83
  three-dimensional version, 86
directed distance, 35
distance, 99–111
  city-block, 100–102
  directed, 35
  Euclidean, 99–100
  homogeneity of, 100
  non-Euclidean, 100–106
  positivity of, 100
  rotation invariance of, 100
  symmetry of, 99
  translation invariance of, 99
division of inequalities, 21
dual problem, 83

ellipsoid, 90

Euclidean distance, 99–100
experimentation, mathematical, 48

Fermat's principle, 84
forward induction, 55–57

Gauss' mean, 76–78
geometric mean, 48
geometry of n dimensions, 112
Graustein, W. C., 65 (footnote)
"greater than" relationship, 5

harmonic mean, 52
Hölder inequality, 68, 73–74
homogeneity, 100

incidence, angle of, 85
induction, mathematical, 19
    backward, 57–59
    forward, 55–57
inequality, 5–6, 9–10
    addition of, 17
    arithmetic–mean—geometric–mean,
        48–61, 73–74
    Buniakowski, 63 (footnote)
    Cauchy, 62–67, 73–74
    Cauchy-Schwarz, 63 (footnote)
    division of, 21
    graph of, 36–40
    Hölder, 67, 73–74
    Minkowski, 72, 73–74
    mixed, 9
    multiplication of, 19
    multiplication of, by a number, 18
    negation of, 10
    for powers, 22
    for roots, 22
    Schwarz, 63 (footnote)
    for squares, 11
    strict, 9
    subtraction of, 18
    transitivity law for, 16
    triangle, 42–44, 69–71, 73–74

Kazarinoff, N. D., 81, 83

Lagrange, J. L., 66
"less than" relationship, 9

magnitude, 25
mathematical induction, 19
    backward, 57–59
    forward, 55–57
maximization, 79ff
maximum function, 27
mean
    arithmetic, 48
    arithmetic-geometric, 76–78
    of Gauss, 76–78
    geometric, 48
    harmonic, 52
    root-mean-square, 61
minimization, 79ff
minimum function, 28
Minkowski inequality, 72, 73–74
mixed inequality, 9
motion, rigid-body, 106
multiplication of inequalities, 19
multiplication of inequality by a
    number, 18

negation of inequality, 10
negative number, 7
    product involving a, 10
negative of a number, 7
Niven, Ivan, 12, 61
non-Euclidean distance, 100–106
number
    negative, 10
    negative of a, 10
    negative, product involving a, 10
    pairing of, 7–8
    positive, 6
    reciprocal of a, 22

order, 12 (footnote)
    complete, 12 (footnote)
ordinate, 35
Osgood, W. F., 65 (footnote)

pairing of numbers, 7–8
perimeter, 87–88
point set, 106
positive number, 6
powers, 22
product, 10–11
  involving a negative number, 10
Pythagorean relationship, 40, 41

ray of light, 83–86
reciprocal, 22
reflection, angle of, 85
refraction, Snell's law of, 86
reverse problem, 83
rigid-body motion, 106
rise, 35
root-mean-square, 61
roots, 22
rotation invariance, 100
run, 35

Schwarz, H. A., 63 (footnote)
set
  convex, 107
  symmetric, 107
sign function, 35–36

slope, 35
  average, 35
  left-hand, 35
  right-hand, 35
Snell's law of refraction, 86
specialization, 58
squares, inequality for, 11
square roots, 40–41
strict inequality, 9
subtraction of inequalities, 18
symmetric means, 75
symmetric point set, 107
symmetry, 99

tangent, 93–97
transitivity, 16
translation invariance, 99
triangle inequality, 42–44, 69–71,
  73–74

unit circle, 102–106
  exterior, 106
  interior, 106
unit disc, 106
  boundary of, 106
upper bound, 12 (footnote)

zero, 6